Data Analytics

CHAPMAN & HALL/CRC DATA SCIENCE SERIES

Reflecting the interdisciplinary nature of the field, this book series brings together researchers, practitioners, and instructors from statistics, computer science, machine learning, and analytics. The series will publish cutting-edge research, industry applications, and textbooks in data science.

The inclusion of concrete examples, applications, and methods is highly encouraged. The scope of the series includes titles in the areas of machine learning, pattern recognition, predictive analytics, business analytics, Big Data, visualization, programming, software, learning analytics, data wrangling, interactive graphics, and reproducible research.

Published Titles

Public Policy Analytics
Code and Context for Data Science in Government
Ken Steif

Feature Engineering and Selection
A Practical Approach for Predictive Models
Max Kuhn and Kjell Johnson

Probability and Statistics for Data Science
Math + R + Data
Norman Matloff

Introduction to Data Science
Data Analysis and Prediction Algorithms with R
Rafael A. Irizarry

Cybersecurity Analytics
Rakesh M. Verma and David J. Marchette

Basketball Data Science
With Applications in R
Paola Zuccolotto and Marcia Manisera

JavaScript for Data Science
Maya Gans, Toby Hodges, and Greg Wilson

Statistical Foundations of Data Science
Jianqing Fan, Runze Li, Cun-Hui Zhang, and Hui Zou

Explanatory Model Analysis
Explore, Explain, and, Examine Predictive Models
Przemyslaw Biecek, Tomasz Burzykowski

Data Analytics
A Small Data Approach
Shuai Huang and Houtao Deng

For more information about this series, please visit: https://www.routledge.com/Chapman--HallCRC-Data-Science-Series/book-series/CHDSS

Data Analytics
A Small Data Approach

Shuai Huang
Houtao Deng

CRC Press
Taylor & Francis Group
Boca Raton London New York

CRC Press is an imprint of the
Taylor & Francis Group, an **informa** business

A CHAPMAN & HALL BOOK

First edition published 2021
by CRC Press
6000 Broken Sound Parkway NW, Suite 300, Boca Raton, FL 33487-2742

and by CRC Press
2 Park Square, Milton Park, Abingdon, Oxon, OX14 4RN

© 2021 Taylor & Francis Group, LLC

CRC Press is an imprint of Taylor & Francis Group, LLC

Library of Congress Cataloging-in-Publication Data

Names: Huang, Shuai (Industrial engineer), author. | Deng, Houtao, author.
Title: Data analytics : a small data approach / Shuai Huang & Houtao Deng.
Description: First edition. | Boca Raton : CRC Press, 2021. | Includes
index. | Summary: "Data Analytics: A Small Data Approach is suitable for
an introductory data analytics course to help students understand some
main statistical learning models. It has many small datasets to guide
students to work out pencil solutions of the models and then compare
with results obtained from established R packages. Also, as data science
practice is a process that should be told as a story, in this book there
are many course materials about exploratory data analysis, residual
analysis, and flowcharts to develop and validate models and data
pipelines. The main models covered in this book include linear
regression, logistic regression, tree models and random forests,
ensemble learning, sparse learning, principal component analysis, kernel
methods including the support vector machine and kernel regression, and
deep learning. Each chapter introduces two or three techniques. For each
technique, the book highlights the intuition and rationale first, then
shows how mathematics is used to articulate the intuition and formulate
the learning problem. R is used to implement the techniques on both
simulated and real-world dataset. Python code is also available at the
book's website: http://dataanalyticsbook.info. Shuai Huang is an
associate professor at the department of industrial & systems
engineering at the university of Washington. He conducts
interdisciplinary research in machine learning, data analytics, and
applied operations research with applications on healthcare,
manufacturing, and transportation areas. Houtao Deng is a data science
researcher and practitioner. He developed several new decision tree
methods such as inTrees. He has built data-driven products for
forecasting, scheduling, pricing, recommendation, fraud detection, and
image recognition"-- Provided by publisher.
Identifiers: LCCN 2021001775 (print) | LCCN 2021001776 (ebook) | ISBN
9780367609504 (hardback) | ISBN 9781003102656 (ebook)
Subjects: LCSH: Quantitative research. | Quantitative research--Data
processing. | R (Computer program language) | Python (Computer program language)
Classification: LCC QA76.9.Q36 H83 2021 (print) | LCC QA76.9.Q36 (ebook)
| DDC 001.4/2--dc23
LC record available at https://lccn.loc.gov/2021001775
LC ebook record available at https://lccn.loc.gov/2021001776

ISBN: 9780367609504 (hbk)
ISBN: 9780367609511 (pbk)
ISBN: 9781003102656 (ebk)

Typeset in URWPalladioL
by KnowledgeWorks Global Ltd.

"We know $\underbrace{\text{the sound of two hands clapping}}_{\textit{Data}}$.

But what is $\underbrace{\text{the sound of one hand clapping?}}_{\textit{Analytics}}$"

J.D. Salinger, *Nine Stories*

We know the sound of two hands clapping.

But what is the sound of one hand clapping?

To Our Families.

To Our Families

Contents

Preface

This book is suitable for an introductory course of data analytics
to help students understand some main statistical learning models,
such as linear regression, logistic regression, tree models and random
forests, ensemble learning, sparse learning, principal component
analysis, kernel methods including the support vector machine and
kernel regression, etc. Data science practice is a process that should
be told as a story, rather than a one-time implementation of one
single model. This process is a main focus of this book, with many
course materials about exploratory data analysis, residual analysis,
and flowcharts to develop and validate models and data pipelines.

There are 10 chapters. Except for Chapter 1, which gives an
overview of the book, each chapter will introduce two or three techniques. For each technique, we will highlight the intuition and rationale behind it. We then articulate the intuition, use math to formulate
the learning problem, and present the full version of the analytic formulation. We use R to implement the technique on both simulated
or real-world datasets, present the analysis process (together with R
code), show the dynamics in the analysis process, and comment on
the results. Some Remarks are also made at the end of each chapter
to enhance understanding of the techniques, reveal their different
natures by other perspectives, reveal their limitations, and mention
existing remedies to overcome these limitations.

There are three unique aspects to this book.

First, instructors will find many small datasets (i.e., consisting of
5—10 data points of 2—4 variables) in this book for models to be
manually implemented by students using step-by-step process. The
idea is to let students work out pencil solutions and then compare
them with results obtained from established R packages. For example, a dataset with 3 data points and 2 predictors is used to illustrate
how the shooting algorithm of LASSO could be implemented both
on paper and in the R package `glmnet`. Another example is that,
to understand the concept of the support vector machine (SVM), we
use a dataset with 4 data points and 2 predictors to illustrate how the
dual formulation of SVM could be solved manually. Furthermore, by

this small dataset we help students see the connection between the computational algorithm with the geometric pattern of the data, i.e., the correspondence between the numeric solution with the so-called support vectors clearly visible in the scatterplot of the data.

Second, instructors will find graphical illustrations to explain some methods to students. These angles exploit connections between the methods; for example, the SVM is illustrated as a neural network; the kernel regression is introduced as a departure from the mindset of global models; and the logistic regression model is introduced as a few creative twists of the modeling process to apply the linear method for a binary classification problem, etc. On a larger scale, the connection between classic statistical models with machine learning algorithms is illustrated by focusing on the understanding of the iterative nature of the computational algorithms enabled by computers. We help students develop an eye for a method's connection with other models that only appear to be different. This understanding will help us know a method's strength and limitations, the importance of the context, and the assumptions we have carried in our data analysis.

Third, it is important for students to understand the storytelling component of data science. Data scientists tell stories every day. A story conveys a message, and a skillful data scientist must have the experience that the message changes its shape and meaning, depending on which model is used, how the model is tuned, or what part of the data is used. And some models have assumed a particular storytelling mode or structure. For example, we found hypothesis testing is a difficult concept for students to understand its essence, because it is a "negative" reading of data. It is not to translate what the data says, but to seek evidence from data against the null hypothesis we will need to come up with first. Examples as such will be found in the book to help students have a larger and deeper view of what they will learn.

Acknowledgments

The first draft of this book was written in the summer of 2017 to be used as the textbook for a new course about Data Analytics (IND E 498) in the Department of Industrial & Systems Engineering of the University of Washington-Seattle. The course participants were mostly senior undergraduate students and first-year graduate students who provided invaluable comments and feedback to improve the book. The authors also thank Ameer Hamza Shakur, Jingshuo Feng, Prof. Xiangyu Chang and his students for their generous help on some figures, R code, and a range of R/LaTex tools. We also thank the Alzheimer's Disease Neuroimaging Initiative (ADNI, `https://adni.loni.usc.edu/`) for the data used in this book.

In writing this book, we owe great debt to many people who generously share their materials and codes online. During the three-year writing process, we tried our best to acknowledge and cite all the specific resources we have used, and we may still have missed a few. In online communities such as `GitHub.com` and `stackoverflow.com` and numerous personal websites/blogs, you can find free resources which can help you quickly start a new project. Most importantly, this book in its current form wouldn't be possible without R and RStudio (`https://www.rstudio.com`), `bookdown` (`https://bookdown.org/`)[1], and the Tufte-LaTeX Developers.

Last, but not least, the authors would like to take this opportunity to thank their editor, John Kimmel, for his support and encouragement throughout the development of this book. The authors also would like to thank the anonymous reviewers who have given great comments and the project editor Michele Dimont and the copyeditor's remarkable work to improve the book.

[1] Xie, Y., *Bookdown: Authoring Books and Technical Documents with R Markdown*, CRC Press, 2019.

Chapter 1: Introduction

Who will benefit from this book?

Students who will find this book useful are those who have not systematically learned statistics or machine learning, but have had some exposure to basic statistical knowledge such as normal distribution, hypothesis testing, and are interested in finding data scientist jobs in a variety of areas. And practitioners with or without formal training in data science-related disciplines, who use data science in interdisciplinary areas, will find this book a useful addition. For example, I know a friend who learned statistics in college, more or less as applied mathematics that less emphasized data, computation, and storytelling, had found a remarkable resemblance between many data science methods with some concepts that she learned 20 years ago. She said if she could have a book that helps connect all the dots, and go through the materials with an easy-to-follow programming tool, it would help her move to a new field of work, as she is a physicist who is now working on genetics data.

We feel the same way. We have been working with medical doctors to diagnose surgical site infections using mobile phone images, with healthcare professionals to use hospital data to optimize the care process, with biologists and epidemiologists to understand the natural history of diseases, and with manufacturing companies to build Internet-of-Things, among others. The challenge of interdisciplinary collaboration is to cross boundaries and build new platforms. To embark on this adventure, a flexible understanding of our methods is important, as well as the skill of storytelling.

To help readers develop these skills, the style of the book highlights a combination of two aspects: technical concreteness and holistic thinking[2]. Holistic thinking is the foundation of how we formulate problems and why we could trust our formulations, knowing that our formulations inevitably are only a partial representation of a real-world problem. Holistic thinking is also the foundation of communication between team members of different backgrounds. With a diverse team, things that make sense intuitively are important to

[2] The chapters are named using different qualities of holistic thinking in decision-makings, including "Abstraction", "Recognition", "Resonance", "Learning", "Diagnosis", "Scalability", "Pragmatism", and "Synthesis".

build team-wide trust in decision-making. And technical concreteness is the springboard for us to jump into a higher awareness and understanding of the problem to make holistic decisions.

To begin our journey, first, let's look at the big picture, the data analytics pipeline.

Overview of a data analytics pipeline

A typical data analytics pipeline consists of several major pillars. In the example shown in Figure 1, it has four pillars: sensor and devices, data preprocessing and feature engineering, feature selection and dimension reduction, and modeling and data analysis. While this is not the only way to present the diverse data pipelines in the real world, these pipelines more or less resemble this architecture.

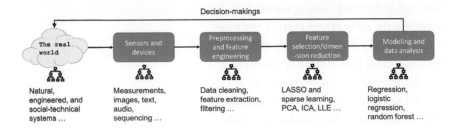

Figure 1: Overview of a data analytics pipeline

The pipeline starts with a real-world problem, for which we are not sure about the underlying system/mechanism, but we are able to characterize the system by defining some variables. Then, we could develop sensors and devices to acquire measurements of these variables[3]. Before analyzing the data and building models, there is a step for data preprocessing and feature engineering. For example, some signals acquired by sensors are not interpretable or not easily compatible with human perceptions, such as the signal acquired by MRI scanning machines in the Fourier space. Data preprocessing also refers to removal of outliers or imputation of missing data, detection and removal of redundant features, to name a few. After data preprocessing, we may conduct feature selection and dimension reduction to distill or condense signals in the data and reduce noise. Finally, we conduct modeling and data analysis on the prepared dataset to gain knowledge and build models of the real-world system[4].

This book focuses on the last two pillars of this pipeline, the modeling, data analysis, feature selection, and dimension reduction methods. But it is helpful to keep in mind the big picture of a data analytics pipeline. Because in practice, it takes a whole pipeline to make things work.

[3] These measurements, we call data, are objective evidences that we can use to explore the statistical principles or mechanistic laws regulating the system behaviors.

[4] Prediction models are quite common, but other models for decision-makings, such as system modeling, monitoring, intervention, and control, can be built as well.

Topics in a nutshell

Specific techniques that will be introduced in this book are shown below.

Data models (i.e., regression-based techniques)

- Chapter 2: Linear regression, least squares estimation, hypothesis testing, R-squared, First Derivative Test, connection with experimental design, data-generating mechanism, history of adventures in understanding errors, exploratory data analysis (EDA)

- Chapter 3: Logistic regression, generalized least squares estimation, iterative reweighted least squares (IRLS) algorithm, ranking (formulated as a regression problem)

- Chapter 4: Bootstrap, data resampling, nonparametric hypothesis testing, nonparametric confidence interval estimation

- Chapter 5: Overfitting and underfitting, limitation of R-squared, training dataset and testing dataset, random sampling, K-fold cross-validation, the confusion matrix, false positive and false negative, the Receiver Operating Characteristics (ROC) curve, the law of errors, how data scientists work with clients

- Chapter 6: Residual analysis, normal Q-Q plot, Cook's distance, leverage, multicollinearity, heterogeneity, clustering, Gaussian mixture model (GMM), the Expectation-Maximization (EM) algorithm, Jensen Inequality

- Chapter 7: Support Vector Machine (SVM), generalize data versus memorize data, maximum margin, support vectors, model complexity and regularization, primal-dual formulation, quadratic programming, KKT condition, kernel trick, kernel machines, SVM as a neural network model

- Chapter 8: LASSO, sparse learning, L_1-norm and L_2-norm regularization, Ridge regression, feature selection, shooting algorithm, Principal Component Analysis (PCA), eigenvalue decomposition, scree plot

- Chapter 9: Kernel regression as generalization of linear regression model, local smoother regression model, k-nearest neighbor (KNN) regression model, conditional variance regression model, heteroscedasticity, weighted least squares estimation, model extension and stacking

- Chapter 10: Deep learning, neural network, activation function, model primitives, convolution, max pooling, convolutional neural network (CNN)

Algorithmic models (i.e., tree-based techniques)

- Chapter 2: Decision tree, entropy, information gain (IG), node splitting, pre- and post-pruning, empirical error, generalization error, pessimistic error by binomial approximation, greedy recursive splitting

- Chapter 3: System monitoring reformulated as classification problem, real-time contrasts method (RTC), design of monitoring statistics, sliding window, anomaly detection, false alarm

- Chapter 4: Random forest, Gini index, weak classifiers, the probabilistic mechanism about why random forests can create a strong classifier out of many weak classifiers, importance score, partial dependency plot

- Chapter 5: Out-of-bag (OOB) error in random forest

- Chapter 6: Residual analysis, clustering by random forests

- Chapter 7: Ensemble learning, Adaboost, analysis of ensemble learning from statistical, computational, and representational perspectives

- Chapter 10: Automations of pipelines, integration of tree models, feature selection, and regression models in 'inTrees', random forest as a rule generator, rule extraction, pruning, selection, and summarization, confidence and support of rules, variable interactions, rule-based prediction

In this book, we will use lower case letters, e.g., x, to represent scalars, bold face, lower case letters, e.g., x, to represent vectors, and bold face, upper case letters, e.g., X, to represent matrices.

Chapter 2: Abstraction
Regression & Tree Models

Overview

Chapter 2 is about *Abstraction*. It concerns how we model and formulate a problem using *specific mathematical models*. Abstraction is powerful. It begins with identification of a few main entities from the problem, and continues to characterize their relationships. Then we focus on the study of these interconnected entities as a pure mathematical system. Consequences can be analytically established within this abstracted framework, while a phenomenon in a concerned context could be identified as special instances, or manifestations, of the abstracted model. In other words, by making abstraction of a real-world problem, we free ourselves from the application context that is usually unbounded and not well defined.

People often adopt a blackbox view of a real-world problem, as shown in Figure 2. There is one (or more) key performance metrics of the system, called the output variable[5], and there is a set of input variables[6] that may help us predict the output variable. These variables are the *few main entities* identified from the problem, and how the input variables impact the output variable is *one* main type of relationship we develop models to characterize.

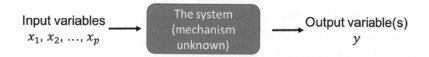

Input variables
$x_1, x_2, ..., x_p$ → The system (mechanism unknown) → Output variable(s) y

[5] Denoted as y, e.g., the yield of a chemical process, the mortality rate of an ICU, the GDP of a nation, etc.
[6] Denoted as $x_1, x_2, ..., x_p$; also called predictors, covariates, features, and, sometimes, factors.

Figure 2: The blackbox nature of many data science problems

These relationships are usually unknown, due to our lack of understanding of the system. It is not always plausible or economically feasible to develop a Newtonian style characterization of the system[7]. Thus, statistical models are needed. They collect data from this blackbox system and build models to characterize the relationship between the input variables and the output variable. Generally, there are two

[7] I.e., using differential equations.

cultures for statistical modeling[8]: One is the **data modeling** culture, while another is the **algorithmic modeling** culture. Linear regression models are examples of the *data modeling* culture; decision tree models are examples of the *algorithmic modeling* culture.

[8] Breiman, L., *Statistical Modeling: The Two Cultures*, Statistical Science, Volume 16, Issue 3, 199-231, 2001.

Two goals are shared by the two cultures: (1) to understand the relationships between the predictors and the output, and (2) to predict the output based on the predictors. The two also share some common criteria to evaluate the success of their models, such as the prediction performance. Another commonality they share is a generic form of their models

$$y = f(x) + \epsilon, \tag{1}$$

where $f(x)$ reflects the *signal* part of y that can be ascertained by knowing x, and ϵ reflects the *noise* part of y that remains uncertain even when we know x. To better illustrate this, we could annotate the model form in Eq. 1 as[9]

$$\underbrace{y}_{data} = \underbrace{f(x)}_{signal} + \underbrace{\epsilon}_{noise} \tag{2}$$

[9] An interesting book about the antagonism between signal and noise: Silver, N., *The Signal and the Noise: Why So Many Predictions Fail–but Some Don't*, Penguin Books, 2015. The author's prediction model, however, failed to predict Donald Trump's victory of the 2016 US Election.

The two cultures differ in their ideas about how to model these two parts. A brief illustration is shown in Table 1.

	$f(x)$	ϵ	"Ideology"
Data Modeling	Explicit form (e.g., linear regression).	Statistical distribution (e.g., Gaussian).	Imply *Cause* and *effect*; uncertainty has a structure.
Algorithmic Modeling	Implicit form (e.g., tree model).	Rarely modeled as structured uncertainty; taken as meaningless noise.	More focus on prediction; to *fit* data rather than to *explain* data.

Table 1: Comparison between the two cultures of models

An illustration of the *data modeling*, using linear regression model, is shown in Figure 3. To develop such a model, we need efforts in two endeavors: the modeling of the signal, and the modeling of the noise (also called errors). It was probably the modeling of the errors, rather than the modeling of the signal, that eventually established a science: Statistics[10].

One only needs to take a look at the beautiful form of the normal distribution (and notice the name as well) to have an impression of its grand status as the law of errors. Comparing with other candidate forms that historically were its competitors, this concentrated, symmetrical, round and smooth form seems a more rational form that a law should take, i.e., see Figure 4.

[10] Errors, as the name suggests, are embarrassment to a theory that claims to be rational. Errors are irrational, like a crack on the smooth surface of rationality. But rationally, if we could find *a law of errors*, we then find the law of irrationality. With that, once again rationality trumps irrationality, and the crack is sealed.

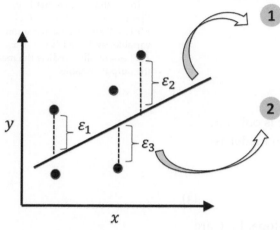

1 *The math form of the signal?*

$$f(x) = \beta_0 + \beta_1 x$$

2 *The law of errors?*

$$\epsilon \sim N(0, \sigma_\varepsilon^2)$$

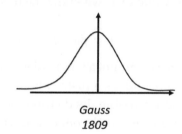

Figure 3: Illustration of the *ideology* of data modeling, i.e., data is used to calibrate, or, estimate, the parameters of a pre-specified mathematical structure

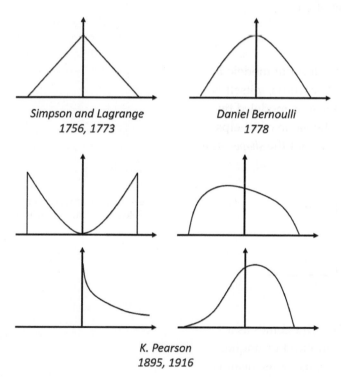

Simpson and Lagrange
1756, 1773

Daniel Bernoulli
1778

Gauss
1809

... shape, (a)symmetry, range ...
The ultimate question: is there one universal law of error?

K. Pearson
1895, 1916

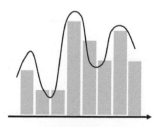

Tschuprow
1910-1916

Figure 4: Hypothesized laws of errors, including the normal distribution (also called the Gaussian distribution, developed by Gauss in 1809) and some of its old rivalries

The ϵ in Eq. 1 is often called the **error term**, noise term, or residual term. ϵ is usually modeled as a Gaussian distribution with mean as 0. The mean has to be 0; otherwise, it contradicts with the name *error*. $f(x)$ is also called the model of the mean structure[11].

Regression models

Rationale and formulation

Let's consider a simple regression model, where there is only one predictor x to predict the outcome y. Linear regression model assumes a linear form of $f(x)$

$$f(x) = \beta_0 + \beta_1 x, \tag{3}$$

where β_0 is called the **intercept**, and β_1 is called the **slope**. Both are also called **regression coefficients**, or more generally, **parameters**.

And ϵ is modeled as a normal distribution[12] with mean 0,

$$\epsilon \sim N\left(0, \sigma_\varepsilon^2\right), \tag{4}$$

where σ_ε^2 is the **variance** of the error.

For any given value of x, we know the model of y is

$$y = \beta_0 + \beta_1 x + \epsilon. \tag{5}$$

As Figure 5 reveals, in linear regression model, y is not modeled as a numerical value, but as a distribution. In other words, y itself is treated as a random variable. Its distribution's mean is modeled by x and the variance is *inherited* from ϵ. Knowing the value of x helps us to determine the *location* of this distribution, but not the *shape*—the shape is always fixed.

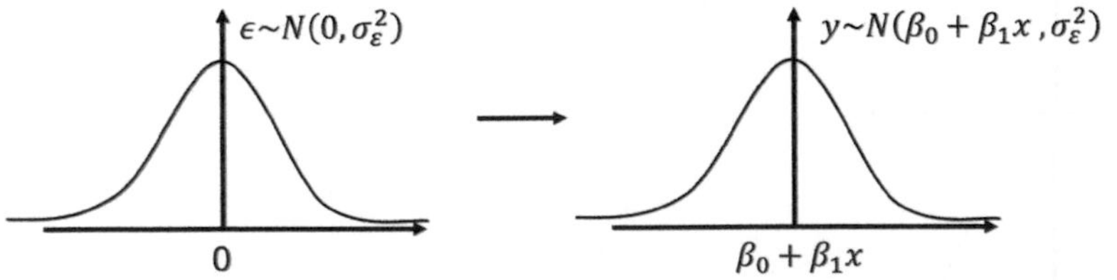

To make a prediction of y for any given x, $\beta_0 + \beta_1 x$ comes as a natural choice. It is too natural that it is often unnoticed or unquestioned. Nonetheless, to predict a random variable, using its mean is the "best" choice, but it is not the only possibility, as Figure 5 reveals that y itself is a random variable, and to predict a random variable, we could also use a confidence interval instead of a point estimate.

[11] To see that, notice that $E(y) = E[f(x) + \epsilon] = E[f(x)] + E[\epsilon]$. Since $E(\epsilon) = 0$ and $f(x)$ is not a random variable, we have $E(y) = f(x)$. Thus, $f(x)$ essentially predicts the mean of the output variable.

[12] I.e., could be other types of distributions, but normal distribution is the norm.

Figure 5: In a linear regression model, y is modeled as a distribution as well

It depends on what you'd like to predict. If the goal is to predict what is the most likely value for y given x, then the best guess is $\beta_0 + \beta_1 x$.[13]

There are more assumptions that have been made to enable the model in Eq. 5.

[13] An important job for statisticians is to prove some ideas are our best choices, i.e., by showing that these choices are optimal decisions under some specific conditions (accurately defined by mathematical terms). It is often that intuitions come before proofs, so many theories are actually developed retrospectively.

- There is a linear relationship between x and y. And this linear relationship remains the same for all the values of x. This is often referred to as a *global* relationship between x and y. Sometimes this assumption is considered strong, e.g., as shown in Figure 6, in drug research it is often found that the dose (x) is related to the effect of the drug (y) in a varying manner that depends on the value of x. Still, from Figure 6 we can see that the linear line captures an essential component in the relationship between x and y, providing a good statistical approximation. Regression models that capture *locality* in the relationship between x and y are introduced in **Chapter 9**.

- The model acknowledges a degree of unpredictability of y. Eq. 5 indicates that y is generated by a combination of the signal (i.e., $\beta_0 + \beta_1 x$) and the noise (i.e., ϵ). Since we could never predict noise, we compute a metric called **R-squared** to quantify the predictability of a model

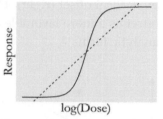

$$\text{R-squared} = \frac{\sigma_y^2 - \sigma_\epsilon^2}{\sigma_y^2}. \qquad (6)$$

Here, σ_y^2 is the variance of y. The *R-squared* ranges from 0 (zero predictability) to 1 (perfect predictability).

Figure 6: Complex relationship between dose (x) and drug response (y), while the linear line does provide a good statistical approximation

- The *significance* of x in predicting y, and the *accuracy* of x in predicting y, are two different concepts. A predictor x could be inadequate in predicting y, i.e., the R-squared could be as low as 0.1, but it still could be statistically significant. In other words, the relation between x and y is not strong, but it is not spurious either. This often happens in social science research and education research projects. Some scenarios are shown in Figure 7.

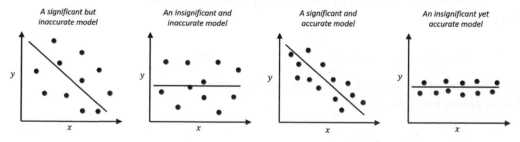

Figure 7: Significance vs. accuracy

- The noise is usually modeled as a normal distribution, but this assumption could be relaxed. A detailed discussion about how to check the normality assumption in data analysis can be found in **Chapter 5**.

Theory and method

Parameter estimation   To estimate a model is to estimate its parameters, i.e., for the model shown in Eq. 5, unknown parameters include β_0, β_1, and σ_ε^2. Usually, we estimate the regression coefficients first. Then, as shown in Figure 3, errors could be computed, and further, σ_ε^2 could be estimated[14].

A training dataset is collected to estimate the parameters. The basic idea is that the best estimate should lead to a line, as shown in Figure 3, that fits the training data as close as possible. To quantify this quality of fitness of a line, two principles are shown in Figure 8: one based on perpendicular offset (left), while another one based on vertical offset (right). History of statistics has chosen the vertical offset as a more favorable approach, since it leads to tractability in analytic forms[15].

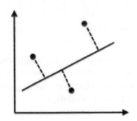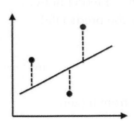

The principle of minimizing vertical offsets leads to the **least-squares estimation** of linear regression models. We can exercise the least squares estimation using the simple regression model shown in Eq. 5. The objective, based on the principle suggested in Figure 8 (right), is to find the line that **minimizes** the **sum of the squared** of the vertical derivations of the observed data points from the line.

Suppose that we have collected N data points, denoted as, (x_n, y_n) for $n = 1, 2, \ldots, N$ [16]. For each data point, i.e., the n_{th} data point, the residual ϵ_n is defined as

$$\epsilon_n = y_n - (\beta_0 + \beta_1 x_n). \tag{7}$$

Then, we define the sum of the squared of the vertical derivations of the observed data points from the line as

$$l(\beta_0, \beta_1) = \sum_{n=1}^{N} \epsilon_n^2. \tag{8}$$

[14] I.e., as a standard practice of sample variance estimation by taking the residuals (i.e., ϵ_1, ϵ_2 and ϵ_3) as *samples* of the population of *error*.

[15] When there were no computers yet, analytic tractability was, and still is, held as a sacred quality of a model.

Figure 8: Two principles to fit a linear regression model: (left) perpendicular offsets; (right) vertical offsets. The distances between the dots (the training data) with the line (the trained model) provide a quantitative metric of how well the model fits the data.

[16] Data is paired, i.e., y_n corresponds to x_n.

Plugging Eq. 7 in Eq. 8 we have

$$l(\beta_0, \beta_1) = \sum_{n=1}^{N} [y_n - (\beta_0 + \beta_1 x_n)]^2. \tag{9}$$

To estimate β_0 and β_1 is to minimize this least squares **loss function** $l(\beta_0, \beta_1)$. This is an **unconstrained continuous optimization** problem. We take derivatives of $l(\beta_0, \beta_1)$ regarding the two parameters and set them to be zero, to derive the estimation equations—this is a common practice of the **First Derivative Test**, illustrated in Figure 9.

$$\frac{\partial l(\beta_0, \beta_1)}{\partial \beta_0} = -2 \sum_{n=1}^{N} [y_n - (\beta_0 + \beta_1 x_n)] = 0,$$

$$\frac{\partial l(\beta_0, \beta_1)}{\partial \beta_1} = -2 \sum_{n=1}^{N} x_n [y_n - (\beta_0 + \beta_1 x_n)] = 0.$$

These two could be rewritten in a more succinct way

$$\begin{bmatrix} N & \sum_{n=1}^{N} x_n \\ \sum_{n=1}^{N} x_n & \sum_{n=1}^{N} x_n^2 \end{bmatrix} \begin{bmatrix} \beta_0 \\ \beta_1 \end{bmatrix} = \begin{bmatrix} \sum_{n=1}^{N} y_n \\ \sum_{n=1}^{N} x_n y_n \end{bmatrix}.$$

We solve these two equations and derive the estimators of β_0 and β_1, denoted as $\hat{\beta}_0$ and $\hat{\beta}_1$, respectively, as

$$\hat{\beta}_1 = \frac{\sum_{n=1}^{N} (x_n - \bar{x})(y_n - \bar{y})}{\sum_{n=1}^{N} x_n^2 - N\bar{x}^2},$$

$$\hat{\beta}_0 = \bar{y} - \hat{\beta}_1 \bar{x}. \tag{10}$$

where $\bar{x}$ and $\bar{y}$ are the sample mean of the two variables, respectively.

There is a structure hidden inside Eq. 10. Note that the estimator $\hat{\beta}_1$ can be rewritten as

$$\hat{\beta}_1 = \frac{\sum_{n=1}^{N} (x_n - \bar{x})(y_n - \bar{y})}{N - 1} \Big/ \frac{\sum_{n=1}^{N} x_n^2 - N\bar{x}^2}{N - 1}, \tag{11}$$

and note that the sample variance of x is defined as

$$\text{var}(x) = \frac{\sum_{n=1}^{N} x_n^2 - N\bar{x}^2}{N - 1},$$

while the numerator in Eq. 11 is called the **sample covariance**[17].

Thus, we can *rewrite* the estimators of β_1 and β_0 as

$$\hat{\beta}_1 = \frac{\text{cov}(x, y)}{\text{var}(x)},$$

$$\hat{\beta}_0 = \bar{y} - \hat{\beta}_1 \bar{x}. \tag{12}$$

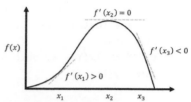

Figure 9: Illustration of the **First Derivative Test** in optimization, i.e., the optimal solution would lead the first derivative to be zero. It is widely used in statistics and machine learning to find optimal solutions of some model formulations. More applications of this technique can be found in later chapters.

[17] The covariance is a measure of the joint variability of two random variables. Denoted as $\text{cov}(x, y)$, the larger the covariance, the stronger the two variables interact.

x	1	3	3	5	5	6	8	9
y	2	3	5	4	6	5	7	8

Table 2: An example dataset

A small data example   Let's practice the estimation method using a simple example. The dataset is shown in Table 2.

Following Eq. 10 we can get $\beta_0 = -1.0714$ and $\beta_1 = 1.2143$. The R codes to verify your calculation are shown below.

```
# Simple example of regression with one predictor
data = data.frame(rbind(c(1,2),c(3,3),c(3,5),
                        c(5,4),c(5,6),c(6,5),
                        c(8,7),c(9,8)))
colnames(data) = c("Y","X")
str(data)
lm.YX <- lm(Y ~ X, data = data)
summary(lm.YX)
```

Extension to multivariate regression model   Consider a more general case where there are more than one predictor

$$y = \beta_0 + \sum_{i=1}^{p} \beta_i x_i + \varepsilon. \tag{13}$$

To fit this multivariate linear regression model with p predictors, we collect N data points, denoted as

$$y = \begin{bmatrix} y_1 \\ y_2 \\ \vdots \\ y_N \end{bmatrix}, \quad X = \begin{bmatrix} 1 & x_{11} & x_{21} & \cdots & x_{p1} \\ 1 & x_{12} & x_{22} & \cdots & x_{p2} \\ \vdots & \vdots & \vdots & \vdots & \vdots \\ 1 & x_{1N} & x_{2N} & \cdots & x_{pN} \end{bmatrix}.$$

where $y \in R^{N \times 1}$ denotes for the N measurements of the outcome variable, and $X \in R^{N \times (p+1)}$ denotes for the data matrix that includes the N measurements of the p input variables and the intercept term, β_0, i.e., the first column of X corresponds to β_0.[18]

To estimate the regression coefficients in Eq. 13, again, we use the least squares estimation method. The first step is to calculate the sum of the squared of the vertical derivations of the observed data points from "the line"[19]. Following Eq. 7, we can define the residual as

[18] Again, the data is paired, i.e., y_n corresponds to x_n that is the n_{th} row of the matrix X.

[19] Here, actually, a hyperplane.

$$\epsilon_n = y_n - \left(\beta_0 + \sum_{i=1}^{p} \beta_i x_{in} \right). \tag{14}$$

Then, following Eq. 8, the sum of the squared of the vertical derivations of the observed data points from "the line" is

$$l(\beta_0, ..., \beta_p) = \sum_{n=1}^{N} \epsilon_n^2. \tag{15}$$

This is again an unconstrained continuous optimization problem, that could be solved by the same procedure we have done for the simple linear regression model. Here, we show how a vector-/matrix-based representation of this derivation process could make things easier.

Let's write up the regression coefficients and residuals in vector forms as

$$\beta = \begin{bmatrix} \beta_0 \\ \beta_1 \\ \vdots \\ \beta_p \end{bmatrix}, \text{ and } \varepsilon = \begin{bmatrix} \varepsilon_1 \\ \varepsilon_2 \\ \vdots \\ \varepsilon_N \end{bmatrix}.$$

Here, $\beta \in R^{(p+1) \times 1}$ denotes for the regression parameters and $\varepsilon \in R^{N \times 1}$ denotes for the N residuals which are assumed to follow a normal distribution with mean as zero and variance as σ_ε^2.

Then, based on Eq. 14, we rewrite ε as

$$\varepsilon = y - X\beta.$$

Eq. 15 could be rewritten as

$$l(\beta) = (y - X\beta)^T (y - X\beta). \tag{16}$$

To estimate β is to solve the optimization problem

$$\min_{\beta} (y - X\beta)^T (y - X\beta).$$

To solve this problem, we can take the gradients of the objective function regarding β and set them to be zero

$$\frac{\partial (y - X\beta)^T (y - X\beta)}{\partial \beta} = 0,$$

which gives rise to the equation

$$X^T (y - X\beta) = 0.$$

This leads to the **least squares estimator** of β as

$$\hat{\beta} = \left(X^T X \right)^{-1} X^T y. \tag{17}$$

A resemblance can be easily detected between the estimator in Eq. 17 with Eq. 12, by noticing that $X^T y$ reflects the correlation[20] between predictors and output, and $X^T X$ reflects the variability[21] of the predictors.

Eq. 17 may come as a surprise to some readers. The regression coefficients, β, by their definition, are supposed to only characterize the

[20] I.e., corresponds to $\text{cov}(x, y)$.

[21] I.e., corresponds to $\text{var}(x)$.

relationship between x and y. However, from Eq. 17, it is clear that the variability of x matters. This is not a contradiction. β and $\widehat{\beta}$ are *two* different entities: β is a theoretical concept, while $\widehat{\beta}$ is a statistical estimate. Statisticians have established theories[22] to study how well $\widehat{\beta}$ estimates β. From Eq. 17, it is clear that where we observe the linear system[23] matters to the modeling of the system. This is one main motivation of the area called the **Design of Experiments** that aims to identify the best locations of x from which we collect observations of the outcome variable, in order to achieve the best parameter estimation results.

[22] E.g, interested readers may read this book: Ravishanker, N. and Dey, D.K., *A First Course in Linear Model Theory*, Chapman & Hall/CRC, 2001.

[23] I.e., from which x we take measurement of y's.

Uncertainty of $\widehat{\beta}$ By generalizing the result in Figure 5 on the multivariate regression, we can see that y is a random vector[24],

[24] "MVN" stands for Multivariate Normal Distribution. See **Appendix** for background knowledge on MVN.

$$y \sim \text{MVN}\left(X^T\beta, \sigma_\varepsilon^2 I\right). \qquad (18)$$

And $\widehat{\beta}$, as shown in Eq. 17, is essentially a *function* of y. Thus, $\widehat{\beta}$ is a random vector as well. In other words, $\widehat{\beta}$ has a distribution. Because of the normality of y, $\widehat{\beta}$ is also distributed as a normal distribution.

The mean of $\widehat{\beta}$ is β, because

$$E(\widehat{\beta}) = E\left[\left(X^TX\right)^{-1}X^Ty\right] = \left(X^TX\right)^{-1}X^TE[y] = \left(X^TX\right)^{-1}X^TX\beta = \beta.$$

And the covariance matrix of $\widehat{\beta}$ is

$$\text{cov}(\widehat{\beta}) = \text{cov}\left[\left(X^TX\right)^{-1}X^Ty\right] = \left(X^TX\right)^{-1}X^T\text{cov}(y)X\left(X^TX\right)^{-1}.$$

Because

$$\text{cov}(y) = \sigma_\varepsilon^2 I,$$

we have

$$\text{cov}(\widehat{\beta}) = \sigma_\varepsilon^2\left(X^TX\right)^{-1}.$$

Thus, we have derived that

$$y \sim \text{MVN}\left(X^T\beta, \sigma_\varepsilon^2 I\right) \Rightarrow \widehat{\beta} \sim \text{MVN}\left[\beta, \sigma_\varepsilon^2\left(X^TX\right)^{-1}\right]. \qquad (19)$$

For each individual parameter β_i, we can infer that

$$\hat{\beta}_i \sim N\left(\beta_i, \frac{\sigma_\varepsilon^2}{x_i^T x_i}\right) \qquad (20)$$

Hypothesis testing of regression parameters Eq. 20 lays the foundation for developing hypothesis testing of the regression parameters.

A hypothesis testing begins with a null hypothesis, e.g.,

$$H_0 : \beta_i = 0.$$

If the null hypothesis is true, then based on Eq. 20, we have

$$\hat{\beta}_i \sim N\left(0, \frac{\sigma_\varepsilon^2}{x_i^T x_i}\right). \tag{21}$$

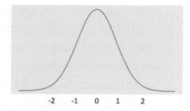

Figure 10: The distribution of $\hat{\beta}_i$

This distribution is shown in Figure 10. It is a graphical display of the possibilities of the values of $\hat{\beta}_i$ that we may observe, *if* H_0 is true.

Then we can derive further implications. Based on Figure 10, we could define a range of $\hat{\beta}_i$ that we believe as most plausible[25]. In other words, if the null hypothesis is true, then it is normal to see $\hat{\beta}_i$ in this range. This thought leads to Figure 11. This is *what is supposed to be*, if the null hypothesis is true. And any value outside of this range is considered as a result of rare chance, noise, or abnormality. We define a level of probability that represents our threshold of rare chance. We coin this threshold level as α.

With the threshold level α, we conclude that any value of $\hat{\beta}_i$ that falls outside of the range is unlikely. If we see $\hat{\beta}_i$ falls outside of the range, we reject the null hypothesis H_0, based on the conflict between *"what is supposed to be"* and *"what happened to be"*.[26] This framework is shown in Figure 11.

Hypothesis testing is a decision made with risks. We may be wrong: even if the null hypothesis is true, there is still a small probability, α, that we may observe $\hat{\beta}_i$ falls outside of the range. But this is not a blind risk. It is a *different kind of risk*: we have scientifically derived the risk, understood it well, and accepted the risk as a cost.

[25] Note that I use the word "plausible" instead of "possible". Any value is always *possible*, according to Eq. 21. But the *possibility* is not equally distributed, as shown in Figure 10. Some values are more possible than others.

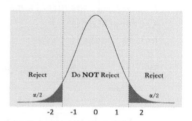

Figure 11: The framework of hypothesis testing

[26] I.e., what we have assumed in H_0 is *what is supposed to be*, and what we have observed in data is *what happened to be*.

R Lab

In this section, we illustrate step-by-step a pipeline of R codes to use the linear regression model in real-world data analysis. Real-world data analysis is challenging. The *real-world* means objectivity, but the *real-worldliness* suggests subjectivity. The purpose of the R codes in this book serves a similar function as a diving coach who dives into the water to show how the action should be done, but the *real-worldliness* can only be felt if you also dive into the water and feel the thrill by yourself. Our data analysis examples try to preserve a certain degree of the *real-worldliness* that embodies both statistical regularities and realistic irregularities[27]. Only the challenge in many real applications is that the boundary between the statistical regularities and realistic irregularities is unclear and undefined.

[27] Prof. George Box once said, *"all models are wrong, some are useful"*.

Having said that, making informed decisions by drawing from rigorous theories, while at the same time, maintaining a critical attitude about theory, are both needed in practices of data analytics.

Here, our data is from a study of Alzheimer's disease[28] that collected some demographics, genetic, and neuroimaging variables from hundreds of subjects. The goal of this dataset is to use these predictors to predict some outcome variables, e.g., one is called the Mini-Mental State Examination (`MMSCORE`), which is a clinical score for determining Alzheimer's disease. It ranges from 1 to 30, while 25 to 30 is normal, 20 to 24 suggests mild dementia, 13 to 20 suggests moderate dementia, and less than 12 indicates severe dementia.

The 5-Step R Pipeline We start with a pipeline of conducting linear regression analysis in R with 5 steps. Please keep in mind that these 5 steps are not a fixed formula: it is a selection of the authors to make it simple.

Step 1 loads the data into the R work environment.

```
# Step 1 -> Read data into R workstation
# RCurl is the R package to read csv file using a link
library(RCurl)
url <- paste0("https://raw.githubusercontent.com",
              "/analyticsbook/book/main/data/AD.csv")
AD <- read.csv(text=getURL(url))
# str(AD)
```

Step 2 is for data preprocessing. This is a standard chunk of code, and it will be used again in future chapters. As this is the first time we see it, here, let's break it into several pieces. The first piece is to create your `X` matrix (predictors) and `Y` vector (outcome variable). The use of `X` for predictors and `Y` for outcome are common practice.

```
# Step 2 -> Data preprocessing.
# Remove variable DX_bl
AD <- AD[ , -which(names(AD) %in% c("DX_bl"))]
# Pick up the first 15 variables for predictors
X <- AD[,1:15]
# Pick up the variable MMSCORE for outcome
Y <- AD$MMSCORE
```

Then, we make a `data.frame` to enclose both the predictors and outcome variable together. Many R functions presume the data are *packaged* in this way.

```
data <- data.frame(X,Y)
names(data)[16] <- c("MMSCORE")
```

[28] Data were obtained from the Alzheimer's Disease Neuroimaging Initiative (ADNI) database (http://adni.loni.usc.edu). The ADNI was launched in 2003 as a public-private partnership, led by Principal Investigator Michael W. Weiner, MD. The primary goal of ADNI has been to test whether serial magnetic resonance imaging (MRI), positron emission tomography (PET), other biological markers, and clinical and neuropsychological assessment can be combined to measure the progression of mild cognitive impairment (MCI) and early Alzheimer's disease (AD).

[29] Usually, there is a client who splits the data for you, sends you the training data only, and withholds the testing data. When you submit your model trained on the training data, the client could verify your model using the testing data. Here, even the dataset we are working on is already the training data, we still split this nominal training data into halves and use one half as the actual training data and the other half as the testing data. Why do we do so? Please see **Chapter 5**.

Then, we split the data into two parts[29]. We name the two parts as *training data* and *testing data*, respectively. The training data is to fit the model. The testing data is excluded from the model training: it will be used to test the model after the final model has been selected using the training data solely.

```
set.seed(1) # generate the same random sequence
# Create a training data (half the original data size)
train.ix <- sample(nrow(data),floor( nrow(data)/2) )
data.train <- data[train.ix,]
# Create a testing data (half the original data size)
data.test <- data[-train.ix,]
```

Step 3 builds up a linear regression model. We use the `lm()` function to fit the regression model[30].

[30] Use `lm()` for more information.

```
# Step 3 -> Use lm() function to build a full
# model with all predictors
lm.AD <- lm(MMSCORE ~ ., data = data.train)
summary(lm.AD)
```

The result is shown in below

```
## Call:
## lm(formula = MMSCORE ~ ., data = data.train)
##
## Residuals:
##     Min      1Q  Median      3Q     Max
## -6.3662 -0.8555  0.1540  1.1241  4.2517
##
## Coefficients:
##              Estimate Std. Error t value Pr(>|t|)
## (Intercept) 17.93920    2.38980   7.507 1.16e-12 ***
## AGE          0.02212    0.01664   1.329 0.185036
## PTGENDER    -0.11141    0.22077  -0.505 0.614280
## PTEDUCAT     0.16943    0.03980   4.257 2.96e-05 ***
## FDG          0.65003    0.17836   3.645 0.000328 ***
## AV45        -1.10136    0.62510  -1.762 0.079348 .
## HippoNV      7.66067    1.68395   4.549 8.52e-06 ***
## e2_1        -0.26059    0.36036  -0.723 0.470291
## e4_1        -0.42123    0.24192  -1.741 0.082925 .
## rs3818361    0.24991    0.21449   1.165 0.245120
## rs744373    -0.25192    0.20787  -1.212 0.226727
## rs11136000  -0.23207    0.21836  -1.063 0.288926
## rs610932    -0.11403    0.21906  -0.521 0.603179
## rs3851179    0.16251    0.21402   0.759 0.448408
## rs3764650    0.47607    0.24428   1.949 0.052470 .
## rs3865444   -0.34550    0.20559  -1.681 0.094149 .
## ---
## Signif. codes:  0 '***' 0.001 '**' 0.01 '*' 0.05 '.' 0.1 ' ' 1
```

```
##
## Residual standard error: 1.63 on 242 degrees of freedom
## Multiple R-squared:  0.3395, Adjusted R-squared:  0.2986
## F-statistic: 8.293 on 15 and 242 DF,  p-value: 3.575e-15
```

Step 4 is model selection. There are many variables that are not significant, i.e., their *p-values* are larger than 0.05. The `step()` function is used for automatic model selection[31], i.e., it implements a brute-force approach to identify the best combinations of variables in a linear regression model.

[31] Use `help(step)` for more information.

```
# Step 4 -> use step() to automatically delete
# all the insignificant variables
# Automatic model selection
lm.AD.reduced <- step(lm.AD, direction="backward", test="F")
```

And the final model the `step()` function identifies is

```
## Step:  AIC=259.92
## MMSCORE ~ PTEDUCAT + FDG + AV45 + HippoNV + e4_1 + rs744373 +
##     rs3764650 + rs3865444
##
##              Df Sum of Sq    RSS    AIC F value   Pr(>F)
## <none>                    658.95 259.92
## - rs744373   1     6.015 664.96 260.27  2.2728 0.132934
## - AV45       1     7.192 666.14 260.72  2.7176 0.100511
## - e4_1       1     8.409 667.36 261.19  3.1774 0.075882 .
## - rs3865444  1     8.428 667.38 261.20  3.1848 0.075544 .
## - rs3764650  1    10.228 669.18 261.90  3.8649 0.050417 .
## - FDG        1    40.285 699.24 273.23 15.2226 0.000123 ***
## - PTEDUCAT   1    44.191 703.14 274.67 16.6988 5.913e-05 ***
## - HippoNV    1    53.445 712.40 278.04 20.1954 1.072e-05 ***
## ---
## Signif. codes:  0 '***' 0.001 '**' 0.01 '*' 0.05 '.' 0.1 ' ' 1
```

It can be seen that the predictors that are kept in the *final model* are all significant. Also, the `R-squared` is 0.3228 using the 8 selected predictors. This is not bad comparing with the `R-squared`, 0.3395, when all the 15 predictors are used (we call this model the *full model*).

We compare the full model with the final model using the F-test that is implemented in `anova()` .

```
anova(lm.AD.reduced,lm.AD)
```

The returned result, shown below, implies that it is statistically indistinguishable between the two models (*p-value* of the F-test is 0.529). The model `lm.AD.reduced` provides an equally good explanation of the data as the full model does, but `lm.AD.reduced` is more economic. The principle of **Occam's razor**[32] would consider the

[32] *"Other things being equal, simpler explanations are generally better than more complex ones"*, is the basic idea of Occam's razor. Albert Einstein was also quoted with a similar expression: *"Everything should be made as simple as possible, but no simpler"*.

model `lm.AD.reduced` more in favor.

```
## Analysis of Variance Table
##
## Model 1: MMSCORE ~ PTEDUCAT + FDG + AV45 + HippoNV +
##     e4_1 + rs744373 + rs3764650 + rs3865444
## Model 2: MMSCORE ~ AGE + PTGENDER + PTEDUCAT + FDG + AV45 +
##     HippoNV + e2_1 + e4_1 + rs3818361 + rs744373 + rs11136000 +
##     rs610932 + rs3851179 + rs3764650 + rs3865444
##   Res.Df    RSS Df Sum of Sq      F Pr(>F)
## 1    249 658.95
## 2    242 642.73  7    16.218 0.8723  0.529
```

Step 5 makes prediction. We can use the function `predict()` [33] which is a function you can find in many R packages. It usually has two main arguments: `obj` is the model, and `data` is the data points you want to predict on. Note that, here, we test the model (that was trained on training data) on the testing data. After gathering the predictions, we use the function `cor()` to measure how close are the predictions with the true outcome values of the testing data. The higher the correlation, the better the predictions.

[33] `predict(obj, data)`

```
# Step 5 -> Predict using your linear regession model
pred.lm <- predict(lm.AD.reduced, data.test)
# For regression model, you can use correlation to measure
# how close your predictions with the true outcome
# values of the data points
cor(pred.lm, data.test$MMSCORE)
```

Beyond the 5-Step Pipeline   The **Exploratory Data Analysis (EDA)** is a practical toolbox that consists of many interesting and insightful methods and tools, mostly empirical and graphical. The idea of EDA was promoted by some statisticians[34]. The EDA could be used before and after we have built the model. For example, a common practice of EDA is to draw the scatterplots to see how potentially the predictors can predict the outcome variable.

[34] E.g., John W. Tukey was a statistician whose career was known to be an advocate of EDA. See his book: *Exploratory Data Analysis*, Addison-Wesley Publishing Co., 1977.

```
# Scatterplot matrix to visualize the relationship
# between outcome variable with continuous predictors
library(ggplot2)
# install.packages("GGally")
library(GGally)
# draw the scatterplots and also empirical
# shapes of the distributions of the variables
p <- ggpairs(AD[,c(16,1,3,4,5,6)],
            upper = list(continuous = "points"),
            lower = list(continuous = "cor"))
print(p)
```

```
library(ggplot2)
qplot(factor(PTGENDER),
      MMSCORE, data = AD,geom=c("boxplot"), fill = factor(PTGENDER))
```

Figure 12 presents the continuous predictors.

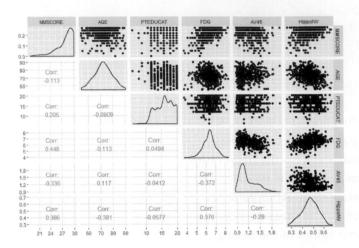

Figure 12: Scatterplots of the continuous predictors versus outcome variable

For the other predictors which are binary, we can use a boxplot, which is shown in Figure 13.

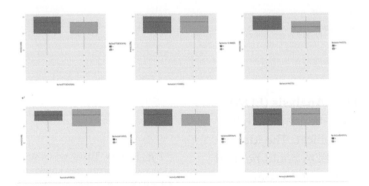

Figure 13: Boxplots of the binary predictors versus outcome variable

In what follows we show another case of EDA.

Consider the relationship between `MMSCORE` and `PTEDUCAT`, and find a graphical way to investigate if the predictor, `AGE`, mediates the relationship between `MMSCORE` and `PTEDUCAT`. One way to do so is to color the data points in the scatterplot (i.e., the color corresponds to the numerical scale of `AGE`). The following R codes generate Figure 14.

```
# How to detect interaction terms
# by exploratory data analysis (EDA)
require(ggplot2)
p <- ggplot(AD, aes(x = PTEDUCAT, y = MMSCORE))
p <- p + geom_point(aes(colour=AGE), size=2)
```

```
# p <- p + geom_smooth(method = "auto")
p <- p + labs(title="MMSE versus PTEDUCAT")
print(p)
```

It looks like that the relationship between `MMSCORE` and `PTEDUCAT` indeed changes according to different levels of `AGE` . While this is subtle, we change the strategy and draw two more figures, i.e., we draw the same scatterplot on two levels of `AGE` , i.e., `AGE < 60` and `AGE > 80` . The following R codes generate Figure 15.

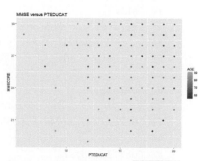

Figure 14: Scatterplots of `MMSCORE` versus `PTEDUCAT`

```
p <- ggplot(AD[which(AD$AGE < 60),],
          aes(x = PTEDUCAT, y = MMSCORE))
p <- p + geom_point(size=2)
p <- p + geom_smooth(method = lm)
p <- p + labs(title="MMSE versus PTEDUCAT when AGE < 60")
print(p)
```

```
p <- ggplot(AD[which(AD$AGE > 80),],
          aes(x = PTEDUCAT, y = MMSCORE))
p <- p + geom_point(size=2)
p <- p + geom_smooth(method = lm)
p <- p + labs(title="MMSE versus PTEDUCAT when AGE > 80")
print(p)
```

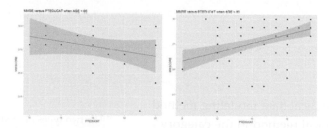

Figure 15: Scatterplots of `MMSCORE` versus `PTEDUCAT` when (left) `AGE < 60` or (right) `AGE > 80`

Figure 15 shows that the relationship between `MMSCORE` and `PTEDUCAT` changes dramatically according to different levels of `AGE` . In other words, it means that the way the predictor `PTEDUCAT` impacts the outcome `MMSCORE` is not simply additive as a regular linear regression model would suggest. Rather, the relationship between the two is modified by `AGE` . This discovery suggests a different mechanism underlying the three variables, as demonstrated in Figure 16.

We then add an interaction term into the regression model

```
# fit the multiple linear regression model
# with an interaction term: AGE*PTEDUCAT
lm.AD.int <- lm(MMSCORE ~ AGE + PTGENDER + PTEDUCAT
                + AGE*PTEDUCAT, data = AD)
summary(lm.AD.int)
```

We can see that this interaction term is significant.

```
##
## Call:
## lm(formula = MMSCORE ~ AGE + PTGENDER
##      + PTEDUCAT + AGE * PTEDUCAT,
##      data = AD)
##
## Residuals:
##     Min      1Q  Median      3Q     Max
## -8.2571 -0.9204  0.5156  1.4219  4.2975
##
## Coefficients:
##               Estimate Std. Error t value Pr(>|t|)
## (Intercept)  40.809411   5.500441   7.419 4.93e-13 ***
## AGE          -0.202043   0.074087  -2.727  0.00661 **
## PTGENDER     -0.470951   0.187143  -2.517  0.01216 *
## PTEDUCAT     -0.642352   0.336212  -1.911  0.05662 .
## AGE:PTEDUCAT  0.011083   0.004557   2.432  0.01534 *
## ---
## Signif. codes:  0 '***' 0.001 '**' 0.01 '*' 0.05 '.' 0.1 ' ' 1
##
## Residual standard error: 2.052 on 512 degrees of freedom
## Multiple R-squared:  0.07193,    Adjusted R-squared:  0.06468
## F-statistic:  9.92 on 4 and 512 DF,  p-value: 9.748e-08
```

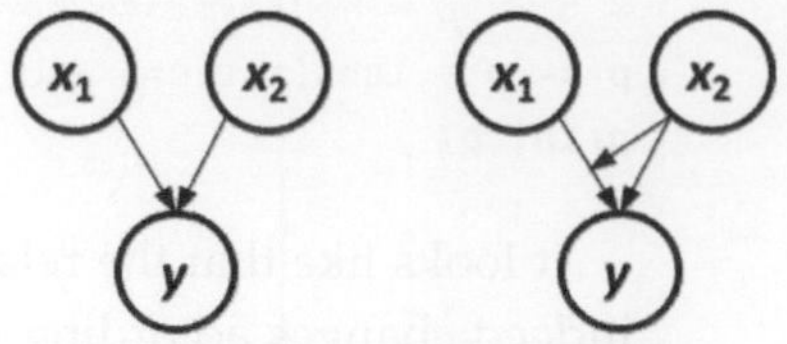

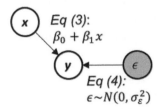

Figure 16: Different data-generating mechanisms: (left) additive relationships between predictors and outcome; (right) additive relationships and interaction

Tree models

Rationale and formulation

While the linear regression model is a typical data modeling method, the decision tree model represents a typical method in the category of algorithmic modeling[35]. The linear regression model, given its many origins and implications, builds a model based on a mathematical characterization of the *data-generating mechanism*, which emphasizes an analytic understanding of the underlying system and how the data is generated from this system[36]. This pursuit of "mechanism" is sometimes too much to ask for if we know little about the physics but only the data, since understanding the mechanism of a problem needs experimental science and profound insights. And this pursuit of "mechanism" limits the applicability of a data modeling method when the data don't seem to follow the data-generating mechanism *prescribed* by the model.

For example, Table 3 shows a dataset that has 6 observations, with two predictors, *Weather* and *Day of week (Dow)*, and an outcome variable, *Play*. Assume that this is a dataset collected by a causal dog walker whose routine includes a sports field.

[35] The two types of modeling cultures are discussed in Table 1.

[36] I.e., Eq. 3 explains how y is impacted by x, and Eq. 4 explains how the rest of y is impacted by a random force. This is illustrated in Figure 17.

Figure 17: The *data-generating mechanism* of a simple linear regression model

ID	Weather	Dow (day of weak)	Play
1	Rainy	Saturday	No
2	Sunny	Saturday	Yes
3	Windy	Tuesday	No
4	Sunny	Saturday	Yes
5	Sunny	Monday	No
6	Windy	Saturday	No

Table 3: Example of a dataset where a decision tree has a home game

It is hard to imagine that, for this dataset, how we can denote the two predictors as x_1 and x_2 and connect it with the outcome variable y in the form of Eq. 13, i.e.,

$$Yes = \beta_0 + \beta_1 Rainy + \beta_2 Tuesday + \epsilon?$$

For this dataset, decision tree is a natural fit. As shown in Figure 18, a decision tree contains a **root node**, **inner nodes**, and **decision nodes** (i.e., the shaded leaf nodes of the tree in Figure 18). For any data point to reach its prediction, it starts from the root node, follows the **splitting rules** alongside the arcs to travel through inner nodes, then finally reaches a decision node. For example, consider the data point *"Weather = Sunny, Dow = Saturday"*, it starts with the root node, *"Weather = Sunny?"*, then goes to inner node *"Dow = Saturday?"*, then reaches the decision node as the left child node of the inner node *"Dow = Saturday?"*. So the decision is *"Play = Yes"*.

Compare with data modeling methods that hope to build a characterization of the data-generating mechanism, algorithmic modeling methods such as the decision tree mimic *heuristics* in human reasoning. It is challenging, while unnecessary, to write up a model of algorithmic modeling in mathematical forms as the one shown in Eq. 13. Algorithmic modeling methods are more *semantics-oriented*, and more focused on patterns detection and description.

Theory/Method

Decision trees could be generated by manual inspection of the data. The one shown in Figure 18 could be easily drawn with a few inspection of the 6 data points in Table 3. Automatic algorithms have been developed that can take a dataset as input and generate a decision tree as output. We can see from Figure 18 that a key element of a decision tree is the *splitting rules* that guide a data point to travel through the inner nodes to reach a final decision node (i.e., to reach a decision).

A splitting rule is defined by a variable and the set of values the variable is allowed to take, e.g., in *"Weather = Sunny?"*, *"Weather"* is the variable and *"Sunny"* is the set of value. The variable used for

Figure 18: Example of a decision tree model

splitting is referred to as the **splitting variable**, and the set of value is referred to as the **splitting value**.

We start with the root node. Possible splitting rules are

- *"Weather = Sunny?"*

- *"Dow = Saturday?"*

- *"Dow = Monday?"*

- *"Dow = Tuesday?"*

Each of the splitting rules will lead to a different root node. Two examples are shown in Figure 19. Which one should we use?

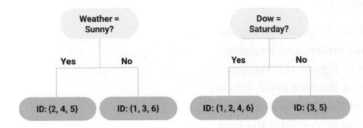

Figure 19: Example of two root nodes

To help us decide on which splitting rule is the best, the concepts **entropy** of data and **information gain (IG)** are needed.

Entropy and information gain (IG)   We can use the concept **entropy** to measure the homogeneity of the data points in a node of the decision tree. It is defined as

$$e = \sum_{i=1,\cdots,K} -P_i \log_2 P_i. \tag{22}$$

where K represents the number of classes of the data points in the node[37], and P_i is the proportion of data points that belong to the class i. The entropy e is defined as zero when the data points in the node all belong to one single class[38]. And $e = 1$ is the maximum value for the entropy of a dataset, i.e., try an example with two classes, where $P_1 = 0.5$ and $P_2 = 0.5$.[39]

A node that consists of data points that are dominated by one class (i.e., entropy is small) is ready to be made a decision node. If it still has a large entropy, splitting it into two child nodes could help reduce the entropy. Thus, to further split a node, we look for the best splitting rule that can maximize the entropy reduction. This entropy reduction can be measured by **IG**, which is the difference of entropy of the parent node and the average entropy of the two child nodes weighted by their number of data points. It is defined as

$$IG = e_s - \sum_{i=1,2} w_i e_i. \tag{23}$$

[37] E.g., in Table 3, there are $K = 2$ classes, *Yes* and *No*.

[38] What is more deterministic than this case?

[39] What is more uncertain than this case?

Here, e_s is the entropy of the parent node, e_i is the entropy of the child node i, and w_i is the number of data points in the child node i divided by the number of data points in the parent node.

For example, for the left tree in Figure 19, using the definition of entropy in Eq. 22, the entropy of the root node is calculated as

$$-\frac{4}{6}\log_2\frac{4}{6} - \frac{2}{6}\log_2\frac{2}{6} = 0.92.$$

The entropy of the left child node ("*Weather = Sunny*") is

$$-\frac{2}{3}\log_2\frac{2}{3} - \frac{1}{3}\log_2\frac{1}{3} = 0.92.$$

The entropy of the right child node ("*Weather != Sunny*") is 0 since all three data points (ID = 1,3,6) belong to the same class.

Then, using the definition of IG in Eq. 23, the IG for the splitting rule "*Weather = Sunny*" is

$$IG = 0.92 - \frac{3}{6} \times 0.92 - \frac{3}{6} \times 0 = 0.46.$$

For the tree in Figure 19 (right), the entropy of the left child node ("*Dow = Saturday*") is

$$-\frac{2}{4}\log_2\frac{2}{4} - \frac{2}{4}\log_2\frac{2}{4} = 1.$$

The entropy of the right child node ("*Dow != Saturday*") is 0 since the two data points (ID = 3,5) belong to the same class.

Thus, the IG for the splitting rule "*Dow = Saturday*" is

$$IF = 0.92 - \frac{4}{6} \times 1 - \frac{2}{6} \times 0 = 0.25.$$

As the IG for the splitting rule "*Weather = Sunny*" is higher, the left tree in Figure 19 is a better choice to start the tree.

Recursive partitioning   The splitting process discussed above could be repeatedly used, until there is no further need to split a node, i.e., the node contains data points from one single class, which is ideal and almost would never happen in reality; or the node has reached the minimum number of data points[40]. This repetitive splitting process is called **recursive partitioning**.

For instance, the left child node in the tree shown in Figure 19 (left) with data points (ID = 2,4,5) still has two classes, and can be further split by selecting the next best splitting rule. The right child node has only one class and becomes a decision node labeled with the decision "*Play = No*".

This greedy approach, like other greedy optimization approaches, is easy to use. One limitation of greedy approches is that they may find **local optimal** solutions instead of **global optimal** solutions. The

[40] It is common to assign a minimum number of data points to prevent the tree-growing algorithm to generate too tiny leaf nodes. This is to prevent "**overfitting**". Elaborated discussion of overfitting will be provided in **Chapter 5**.

optimal choice we made in choosing between the two alternatives in Figure 19 is a *local* optimal choice, and all later nodes of the final tree model are impacted by our decision made on the root node. The *optimal root node* doesn't necessarily lead to the *optimal tree*[41].

An illustration of the risk of getting stuck in a local optimal solution of greedy optimization approaches is shown in Figure 20. Where the algorithm gets started matters to where it ends up. For this reason, decision tree algorithms are often sensitive to data, i.e., it is not uncommon that a slight change of the dataset may cause a considerable change of the topology of the decision tree.

Tree pruning   To enhance the robustness of the decision tree learned by data-driven approaches such as the recursive partitioning, **pruning** methods could be used to cut down some unstable or insignificant branches. There are **pre-pruning** and **post-pruning** methods. Pre-pruning stops growing a tree when a pre-defined criterion is met. For example, one can set the **depth of a tree** (i.e., the depth of a node is the number of edges from the node to the tree's root node; the depth of a tree is the maximum depth of its leaf nodes), or the minimum number of data points at the leaf nodes. These approaches need prior knowledge, and they may not necessarily reflect the characteristics of the particular dataset. More data-dependent approaches can be used. For example, we may set a minimum IG threshold to stop growing a tree when the IG is below the threshold. This may cause another problem, i.e., a small IG at an internal node does not necessarily mean its potential child nodes can only have smaller IG values. Therefore, pre-pruning can cause over-simplified trees and thus **underfitted** tree models. In other words, it may be too cautious.

In contrast, post-pruning prunes a tree after it is fully grown. A fully grown model aggressively spans the tree, i.e., by setting the depth of the tree as a large number. To pursue a fully grown tree is to mitigate the risk of underfit. The cost is that it may overfit the data, so post-pruning is needed. Post-pruning starts from the bottom of the tree. If removing an inner node (together with all the descendant nodes) does not increase the error *significantly*, then it should be pruned. The question is how to evaluate the significance of the increase of error[42].

We will refer readers to **Chapter 5** for understanding more about concepts such as **empirical error** and **generalization error**. Understanding the difference between them is a key step towards maturity in data analytics. Like the difference between *money* and *currency*, the difference will be obvious to you as long as you have seen the difference.

[41] In other words, an optimal tree is the optimal one among all the possible trees, so an optimal root node won't necessarily lead to an optimal tree.

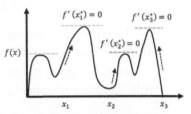

Figure 20: A greedy optimization approach starts its adventure from an **initial solution**. Here, x_1, x_2, x_3 are different initial solutions of 3 usages of the optimization approach, and 3 *optimal* solutions are found, while only one of them is *globally optimal*.

[42] Interested readers may find the discussion in the Remarks section useful.

Extensions and other considerations

- In our data example in Table 3 we only have categorical variables, so candidate splitting rules could be defined relatively easier. For a continuous variable, one approach to identify candidate splitting rules is to order the observed values first, and then, use the average of each pair of consecutive values for splitting.

- If the outcome variable is continuous, we can use the variance of the outcome variable to measure the "entropy" of a node, i.e.,

$$v = \sum_{n=1}^{N} \left(\bar{y} - y_n \right)^2,$$

where $y_{n=1,\cdots,N}$ are the values of the outcome variable in the node, and $\bar{y}$ is the average of the outcome variable. And the information gain can be calculated similarly.

- Both pre-pruning and post-pruning are useful in practices, and it is hard to say which one is better than the other. There is a belief that post-pruning can often outperform pre-pruning. A better procedure is to use **cross-validation**[43]. A popular pre-pruning parameter used in the R package `rpart` is `cp`, i.e., it sets a value such that all splits need to improve the IG by at least a factor of `cp` to be approved. This pre-pruning strategy works well in many applications.

[43] Details are given in **Chapter 5**.

R Lab

The 6-Step R Pipeline   We use `DX_bl` as the outcome variable that is binary[44]. We use other variables (except `ID`, `TOTAL13` and `MMSCORE`) to predict `DX_bl`.

[44] In `DX_bl`, `0` denotes normal subjects; `1` denotes diseased subjects.

   Step 1 loads the needed R packages and data into the workspace.

```
# Key package for decision tree in R:
# rpart (for building the tree);
# rpart.plot (for drawing the tree)
library(RCurl)
library(rpart)
library(rpart.plot)

# Step 1 -> Read data into R workstation
url <- paste0("https://raw.githubusercontent.com",
              "/analyticsbook/book/main/data/AD.csv")
data <- read.csv(text=getURL(url))
```

Step 2 is about data preprocessing.

```
# Step 2 -> Data preprocessing
# Create your X matrix (predictors) and
# Y vector (outcome variable)
X <- data[,2:16]
Y <- data$DX_bl

# The following code makes sure the variable "DX_bl"
# is a "factor".
Y <- paste0("c", Y)
# This line is to "factorize" the variable "DX_bl".
# It denotes "0" as "c0" and "1" as "c1",
# to highlight the fact that
# "DX_bl" is a factor variable, not a numerical variable
Y <- as.factor(Y) # as.factor is to convert any variable
                  # into the format as "factor" variable.

# Then, we integrate everything into a data frame
data <- data.frame(X,Y)
names(data)[16] = c("DX_bl")

set.seed(1) # generate the same random sequence
# Create a training data (half the original data size)
train.ix <- sample(nrow(data),floor( nrow(data)/2) )
data.train <- data[train.ix,]
# Create a testing data (half the original data size)
data.test <- data[-train.ix,]
```

Step 3 is to use the `rpart()` function in the R package `rpart` to build the decision tree.

```
# Step 3 -> use rpart to build the decision tree.
tree <- rpart(DX_bl ~ ., data = data.train)
```

Step 4 is to use the `prp()` function to plot the decision tree[45]

```
# Step 4 -> draw the tree
prp(tree, nn.cex = 1)
```

And the decision tree is shown in Figure 21.

Step 5 is to prune the tree using the R function `prune()`. Remember that the parameter `cp` controls the model complexity[46].

Let us try `cp` = 0.03. This leads to a decision tree as shown in Figure 22.

```
# Step 5 -> prune the tree
tree <- prune(tree,cp=0.03)
prp(tree,nn.cex=1)
```

Step 6 is to evaluate the trained model by predicting the testing data.

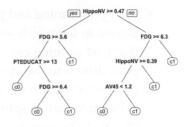

Figure 21: The unpruned decision tree to predict `DX_bl`

[45] `prp()` is a capable function. It has many arguments to specify the details of how the tree should be drawn. Use `help(prp)` to see details.

[46] A larger `cp` leads to a less complex tree.

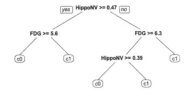

Figure 22: The pruned decision tree model to predict `DX_bl` of the AD data with `cp = 0.03`

```
# Step 6 -> Predict using your tree model
pred.tree <- predict(tree, data.test, type="class")
```

And we can evaluate the prediction performance using error rate.

```
# The following line calculates the prediction error
# rate (a number from 0 to 1) for a binary classification problem
err.tree <- length(which(pred.tree !=
                             data.test$DX_bl))/length(pred.tree)
# 1) which(pred.tree != data$DX_bl) identifies the locations
#    of the incorrect predictions;
# 2) length(any vector) returns the length of that vector;
# 3) thus, the ratio of incorrect prediction over the total
#    prediction is the prediction error
print(err.tree)
```

Remarks

Statistical model vs. causal model

People unconsciously interprets a regression model as a causal model. When an unconscious interpretation is stated, it seems absurd and untrue, but it is fair to say that the line between a statistical model and a causal model is often blurred. We cannot blame ourselves for falling for this temptation before we have had a chance to see it through a critical lens, since both models share the same representation: an asymmetric form where predictors are on one side of the equation and the outcome is on the other side. Plus, the concept of *significance* is no less confusing: a common misinterpretation is to treat the *statistical significance* of a predictor as evidence of *causal significance* in the application context. The fact is that statistical significance doesn't imply that the relationship between the predictor and the outcome variable is causal.

To see this, in what follows we will show an example that the statistical significance of a variable would disappear when some other variables are added into the model. Still using the AD dataset, we fit a regression model using the variable AGE only.

```
lm.AD.age <- lm(MMSCORE ~ AGE, data = AD)
summary(lm.AD.age)
```

And the result is shown below.

```
##
## Call:
## lm(formula = MMSCORE ~  AGE, data = AD)
##
```

```
## Residuals:
##    Min      1Q  Median      3Q     Max
## -8.7020 -0.9653  0.6948  1.6182  2.5447
##
## Coefficients:
##             Estimate Std. Error t value Pr(>|t|)
## (Intercept) 30.44147    0.94564  32.191   <2e-16 ***
## AGE         -0.03333    0.01296  -2.572   0.0104 *
## ---
## Signif. codes:  0 '***' 0.001 '**' 0.01 '*' 0.05 '.' 0.1 ' ' 1
##
## Residual standard error: 2.11 on 515 degrees of freedom
## Multiple R-squared:  0.01268,   Adjusted R-squared:  0.01076
## F-statistic: 6.614 on 1 and 515 DF,  p-value: 0.0104
```

The predictor, `AGE` , is significant since its *p-value* is 0.0104.
Now let's include more demographics variables into the model.

```
# fit the multiple linear regression model
# with more than one predictor
lm.AD.demo <- lm(MMSCORE ~  AGE + PTGENDER + PTEDUCAT,
                 data = AD)
summary(lm.AD.demo)
```

And the result is shown below.

```
##
## Call:
## lm(formula = MMSCORE ~ AGE +
##    PTGENDER + PTEDUCAT, data = AD)
##
## Residuals:
##    Min      1Q  Median      3Q     Max
## -8.4290 -0.9766  0.5796  1.4252  3.4539
##
## Coefficients:
##             Estimate Std. Error t value Pr(>|t|)
## (Intercept) 27.70377    1.11131  24.929  < 2e-16 ***
## AGE         -0.02453    0.01282  -1.913   0.0563 .
## PTGENDER    -0.43356    0.18740  -2.314   0.0211 *
## PTEDUCAT     0.17120    0.03432   4.988 8.35e-07 ***
## ---
## Signif. codes:  0 '***' 0.001 '**' 0.01 '*' 0.05 '.' 0.1 ' ' 1
##
## Residual standard error: 2.062 on 513 degrees of freedom
## Multiple R-squared:  0.0612, Adjusted R-squared:  0.05571
## F-statistic: 11.15 on 3 and 513 DF,  p-value: 4.245e-07
```

Now we can see that the predictor `AGE` is on the boardline of
significance with a *p-value* 0.0563. The other predictors, `PTGENDER`
and `PTEDUCAT` , are significant. The reason that the predictor `AGE`

is now no longer significant is an interesting phenomenon, but it is not unusual in practice that a significant predictor becomes insignificant when other variables are included or excluded[47].

One strategy to mitigate this problem is to explore your data from every possible angle, and try out different model formulations. The goal of your data analysis is not to get a final conclusive model that dictates the rest of the analysis process. The data analysis is an exploratory and dynamic process, i.e., as you see, the dynamic interplay of the variables, how they impact each others' significance in predicting the outcome, is something you could only obtain by analyzing the data in an exploratory and dynamic way. The fact that a model fits the data well and passes the significance test only means that there is nothing significant in the data that is found to be against the model. The goodness-of-fit of the data doesn't mean that the data says this model is the only causal model and other models are impossible.

Design of experiments

Related to this issue of "statistical model vs. causal model", the design of experiments (DOE) is a discipline which provides systematic data collection procedures to render the regression model as a causal model. How this could be done demands a lengthy discussion and illustration[48]. Here, we briefly review its foundation to see why it has the connection with a linear regression model.

We have seen in Eq. 19 that the uncertainty of $\widehat{\beta}$ comes from two sources, the noise in the data that is encoded in σ_ϵ^2, and the structure of X. σ_ϵ^2 reflects essential uncertainty inherent in the system, but X is about how we collect the data. Thus, experimental design methods seek to optimize the structure of X such that the uncertainty of $\widehat{\beta}$ could be minimized.

For example, suppose that there are three predictors. Let's consider the following structure of X

$$X = \begin{bmatrix} 1 & 0 & 0 \\ 0 & 1 & 0 \\ 0 & 0 & 1 \end{bmatrix}.$$

It can be seen that, with this structure, the variance of $\widehat{\beta}$ is[49]

$$cov(\hat{\beta}) = \sigma_\epsilon^2 I_3.$$

In other words, we can draw two main observations. First, the estimations of the regression parameters are now independent, given that their correlations are zero. Second, the variances of the estimated regression parameters are the same. Because of these two traits, this

[47] This is because of the statistical dependence of the estimation of the predictors. Remember that β and $\hat{\beta}$ are two different entities. In the ground truth the two regression coefficients, β_i and β_j, may be independent with each other, but $\hat{\beta}_i$ and $\hat{\beta}_j$ could still be correlated.

As we have known that

$$\text{cov}(\widehat{\beta}) = \sigma_\epsilon^2 \left(X^T X \right)^{-1},$$

as long as $X^T X$ is not an identity matrix, the estimators of the regression parameters are dependent in a complicated and data-dependant way. Due to this reason, we need to be cautious about how to interpret the estimated regression parameters, as they are interrelated constructs.

[48] Interested readers may start with this book: Goos, P. and Jones, B., *Optimal Design of Experiments: A Case Study Approach*, Wiley, 2011.

[49] I is the identity matrix. Here, $I_3 = \begin{bmatrix} 1 & 0 & 0 \\ 0 & 1 & 0 \\ 0 & 0 & 1 \end{bmatrix}$.

data matrix X is ideal and adopted in DOE to create *factorial designs*. For a linear regression model built on a dataset with such a data matrix, adding or deleting variables from the regression model will not result in changes of the estimations of other parameters.

The pessimistic error estimation in post-pruning

Let's look at the tree in Figure 23. It has one root node, one inner node, and three leaf nodes. The target for tree pruning, for this example, is the inner node. In other words, should we prune the inner node and its subsequent child nodes?

We have mentioned that if the improvement on error is not significant, we should prune the node. Let's denote the **empirical error rate**[50] as $\hat{e}$. The reason we give the notation a *hat* is because it is only an estimate of an underlying parameter, the true error e. $\hat{e}$ is usually smaller than e, and thus, it is considered to be optimistic. To create a fairer estimate of e, the **pessimistic error estimation** approach is used for tree pruning.

The pessimistic error estimation, like a regression model, builds on a hypothesized data-generating mechanism. Here, the *data* is the *errors* we observed from the training data. A data point can be either correctly or wrongly classified, and we can view the probability of being wrongly classified as a Bernoulli trial, while the parameter of this Bernoulli trial, commonly denoted as p, is e. If we denote the total number of errors we have observed on the n data points as d, we can derive that d is distributed as a binomial distribution. We can write this data-generating mechanism as

$$d \sim Bino\left(n, e\right).$$

Since n is usually large, we can use the normal approximation for the binomial distribution

$$d \sim N\left(ne, ne(1-e)\right).$$

As $\hat{e} = d/n$, we have

$$\hat{e} \sim N\left(e, \frac{e(1-e)}{n}\right).$$

Skipping further derivations (more assumptions are imposed, indeed, to derive the following conclusion), we can derive the confidence interval of e as

$$\hat{e} - z_{\alpha/2}\sqrt{\frac{\hat{e}(1-\hat{e})}{n}} \le e \le \hat{e} + z_{\alpha/2}\sqrt{\frac{\hat{e}(1-\hat{e})}{n}}.$$

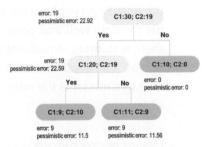

error: 19
pessimistic error: 22.92 C1:30; C2:19

Yes   No

error: 19
pessimistic error: 22.59 C1:20; C2:19   C1:10; C2:0

Yes   No   error: 0
                     pessimistic error: 0

C1:9; C2:10   C1:11; C2:9

error: 9   error: 9
pessimistic error: 11.5   pessimistic error: 11.56

Figure 23: An example of tree pruning using pessimistic error

[50] Empirical error is derived based on the training data.

The upper bound of the interval, $\hat{e} + z_{\alpha/2}\sqrt{\frac{\hat{e}(1-\hat{e})}{n}}$, is named as the *pessimistic error*. The tree pruning methods that use the *pessimistic error* are motivated by a conservative perspective.

The pessimistic error depends on three values: α, which is often set to be 0.25 so that $z_{\alpha/2} = 1.15$; $\hat{e}$, which is the training error rate; and n, which is the number of data points at the node[51].

[51] The pessimistic error is larger with a smaller n, an estimation method that accounts for the sample size.

Now let's revisit Figure 23.

First, let's derive the pessimistic errors for the two child nodes of the inner node. The empirical error rate for the left child node is $\hat{e} = \frac{9}{19} = 0.4737$. For the pessimistic error, we can get that

$$\hat{e} + z_{\alpha/2}\sqrt{\frac{\hat{e}(1-\hat{e})}{n}} = 0.4737 + 1.15\sqrt{\frac{0.4737(1-0.4737)}{19}} = 0.605.$$

With this error rate, for a node with 19 data points, the total misclassified data points can be $mp = 0.605 \times 19 = 11.5$.

For the right child node, the empirical error rate is $\hat{e} = \frac{9}{20} = 0.45$. For the pessimistic error, we can get that

$$\hat{e} + z_{\alpha/2}\sqrt{\frac{\hat{e}(1-\hat{e})}{n}} = 0.45 + 1.15\sqrt{\frac{0.45(1-0.45)}{20}} = 0.578.$$

With this error rate, for a node with 20 data points, the total misclassified data points can be $mp = 0.578 \times 20 = 11.56$.

Thus, if we keep this branch, the total misclassified data points would be $mp = 11.5 + 11.56 = 23.06$.

Now let's evaluate the alternative: to cut the branch. This means the inner node will become a decision node, as shown in Figure 24. We will label the new decision node as C1, since 20 of the included data points are labeled as C1, while 19 are labeled as C2. The empirical error rate e is $\hat{e} = \frac{19}{39} = 0.4871$. For the pessimistic error, we can get that

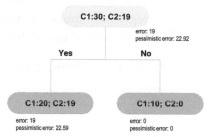

Figure 24: The pruned tree of Figure 23

$$\hat{e} + z_{\alpha/2}\sqrt{\frac{\hat{e}(1-\hat{e})}{n}} = 0.4871 + 1.15\sqrt{\frac{0.4871(1-0.4871)}{39}} = 0.579.$$

With this error rate, for a dataset with 39 data points, the total misclassified data points can be $mp = 0.579 \times 39 = 22.59$. This is what would happen if we prune the tree. As $22.59 < 23.06$, pruning is a better decision.

The pruned tree is shown in Figure 24. A complete post-pruning method will continue to consider further pruning: now consider pruning the child nodes of the root node. Following the process outlined above, the would-be misclassified data points based on the

pessimistic error rate at the root node is 22.92, and the total misclassified instances based on the pessimistic error rate from its child nodes
is $22.59 + 0 = 22.59$. Pruning the child nodes would lead to increased
error. Thus, no further pruning is needed: the child nodes are kept
and the final tree consists of three nodes.

Exercises

ID	x_1	x_2	y
1	−0.15	−0.48	0.46
2	−0.72	−0.54	−0.37
3	1.36	−0.91	−0.27
4	0.61	1.59	1.35
5	−1.11	0.34	−0.11

Table 4: Dataset for building a linear
regression model

1. Here let's consider the dataset in Table 4. Let's build a linear regression model, i.e.,

$$y = \beta_0 + \beta_1 x_1 + \beta_2 x_2 + \epsilon,$$

and

$$\epsilon \sim N\left(0, \sigma_\epsilon^2\right).$$

and calculate the regression parameters $\beta_0, \beta_1, \beta_2$ manually.

2. Follow up the data on Q1. Use the R pipeline to build the linear
regression model. Compare the result from R and the result by
your manual calculation.

3. Read the following output in R.

```
## Call:
## lm(formula = y ~ ., data = data)
##
## Residuals:
##       Min        1Q    Median        3Q       Max
## -0.239169 -0.065621  0.005689  0.064270  0.310456
##
## Coefficients:
##              Estimate Std. Error t value Pr(>|t|)
## (Intercept)  0.009124   0.010473   0.871    0.386
## x1           1.008084   0.008696 115.926   <2e-16 ***
## x2           0.494473   0.009130  54.159   <2e-16 ***
## x3           0.012988   0.010055   1.292    0.200
## x4          -0.002329   0.009422  -0.247    0.805
## ---
## Signif. codes:  0 '***' 0.001 '**' 0.01 '*' 0.05 '.' 0.1 ' ' 1
```

```
##
## Residual standard error: 0.1011 on 95 degrees of freedom
## Multiple R-squared:  0.9942, Adjusted R-squared:  0.994
## F-statistic:  4079 on 4 and 95 DF,  p-value: < 2.2e-16
```

4. (a) Write the fitted regression model. (b) Identify the significant variables. (c) What is the R-squared of this model? Does the model fit the data well? (d) What would you recommend as the next step in data analysis?

5. Consider the dataset in Table 5. Build a decision tree model by manual calculation. To simplify the process, let's only try three alternatives for the splits: $x_1 \geq 0.59$, $x_1 \geq 0.37$, and $x_2 \geq 0.35$.

ID	x_1	x_2	y
1	0.22	0.38	No
2	0.58	0.32	Yes
3	0.57	0.28	Yes
4	0.41	0.43	Yes
5	0.6	0.29	No
6	0.12	0.32	Yes
7	0.25	0.32	Yes
8	0.32	0.38	No

Table 5: Dataset for building a decision tree

6. Follow up on the dataset in Q5. Use the R pipeline for building a decision tree model. Compare the result from R and the result by your manual calculation.

7. Use the `mtcars` dataset in R, select the variable `mpg` as the outcome variable and other variables as predictors, run the R pipeline for linear regression, and summarize your findings.

8. Use the `mtcars` dataset in R, select the variable `mpg` as the outcome variable and other variables as predictors, run the R pipeline for decision tree, and summarize your findings. Another dataset is to use the `iris` dataset, select the variable `Species` as the outcome variable (i.e., to build a classification tree).

9. Design a simulated experiment to evaluate the effectiveness of the `lm()` in R. For instance, you can simulate 100 samples from a linear regression model with 2 variables,

$$y = \beta_1 x_1 + \beta_2 x_2 + \epsilon,$$

where $\beta_1 = 1$, $\beta_2 = 1$, and

$$\epsilon \sim N(0,1).$$

You can simulate x_1 and x_2 using the standard normal distribution $N(0,1)$. Run `lm()` on the simulated data, and see how close the fitted model is with the true model.

10. Follow up on the experiment in Q9. Let's add two more variables x_3 and x_4 into the dataset but still generate 100 samples from a linear regression model from the same underlying model

$$y = \beta_1 x_1 + \beta_2 x_2 + \epsilon,$$

where $\beta_1 = 1$, $\beta_2 = 1$, and

$$\epsilon \sim N(0,1).$$

In other words, x_3 and x_4 are insignificant variables. You can simulate x_1 to x_4 using the standard normal distribution $N(0,1)$. Run `lm()` on the simulated data, and see how close the fitted model is with the true model.

11. Follow up on the experiment in Q10. Run `rpart()` on the simulated data, and see how close the fitted model is with the true model.

12. Design a simulated experiment to evaluate the effectiveness of the `rpart()` in R package `rpart`. For instance, you can simulate 100 samples from a tree model as shown in Figure 25, run `rpart()` on the simulated data, and see how close the fitted model is with the true model.

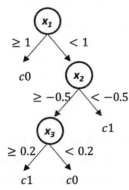

Figure 25: The true model for simulation experiment in Q12

Chapter 3: Recognition
Logistic Regression & Ranking

Overview

Chapter 3 is about *Recognition*. This is an important skill in real-world practices of data analytics. It is to recognize *the same* abstracted form embedded in different real-world problems. No matter how different the problem looks, we hope to leverage existing models and solutions that have been proven effective for the forms that are recognizable in the problem. This is why we say the same model/theory could be applied in multiple areas[52].

This is not to say that a real-world problem is equivalent to an abstracted problem. A dialectic thinking is needed here to understand the relationship between a real-world problem and its reduced form, an abstracted formulation. On one hand, for a real-world problem to be *real-world*, it always has something that exceeds the boundary of a reduced form. On the other hand, for a real-world problem to be solvable, it has to have some kinds of forms.

Many operations researchers believe that being able to recognize these abstracted forms holds the key to solve real-world problems effectively[53]. For some abstracted forms, indeed we have studied them well and are confident to provide a sense of "closure". It takes a sense of closure to conclude that we have solved a real-world problem, or at least we have reached the best solution as far as our knowledge permits. And we have established criteria to evaluate how well we have solved these abstract forms. Those are the territories where we have surveyed in detail and in depth. If to solve a real-world problem is to battle a dragon in its lair, *recognition* is all about paving the way for the dragon to follow the bread crumbs so that we can battle it in a familiar battlefield.

[52] Another practical metaphor is: *a model is a hammer, and applications are nails.*

[53] Some said, *formulation is an art; and a good formulation contributes more than 50% in solving the problem.*

Logistic regression model

Rationale and formulation

Linear regression models are introduced in **Chapter 2** as a tool to predict a continuous response using a few input variables. In some applications, the response variable is a binary variable that denotes two classes. For example, in the AD dataset, we have a variable called `DX_bl` that encodes the diagnosis information of the subjects, i.e., `0` denotes *normal*, while `1` denotes *diseased*.

We have learned about linear regression models to connect the input variables with the outcome variable. It is natural to wonder if the linear regression framework could still be useful here. If we write the regression equation

$$\text{The goal:} \quad \underbrace{y}_{\text{Binary}} \quad = \quad \underbrace{\beta_0 + \sum_{i=1}^{p} \beta_i x_i + \varepsilon.}_{\text{Continuous and unbounded}} \tag{24}$$

Something doesn't make sense. The reason is obvious: the right-hand side of Eq. 24 is continuous without bounds, while the left hand side of the equation is a binary variable. A graphical illustration is shown in Figure 26. So we have to modify this equation, either the right-hand side or the left-hand side.

Since we want it to be a linear model, it is better not to modify the right-hand side. So look at the left-hand side. Why we have to stick with the natural scale of y? We could certainly work out a more linear-model-friendly scale. For example, instead of predicting y, how about predicting the probability $Pr(y = 1|x)$? If we know $Pr(y = 1|x)$, we can certainly convert it to the scale of y.[54]

Thus, we consider the following revised goal

$$\text{Revised goal:} \quad \underbrace{Pr(y = 1|x)}_{\text{Continuous but bounded}} \quad = \quad \underbrace{\beta_0 + \sum_{i=1}^{p} \beta_i x_i + \varepsilon.}_{\text{Continuous and unbounded}} \tag{25}$$

Changing our outcome variable from y to $Pr(y = 1|x)$ is a good move, since $Pr(y = 1|x)$ is on a continuous scale. However, as it is a probability, it has to be in the range of $[0, 1]$. We need more modifications to make things work.

If we make a lot of modifications and things barely work, we may have lost the essence. What is the essence of the linear model that we would like to leverage in this binary prediction problem? Interpretability—sure, the linear form seems easy to understand, but as we have pointed out in **Chapter 2**, this interpretability comes

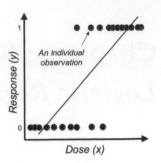

Figure 26: Direct application of linear regression on binary outcome, i.e., illustration of Eq. 24 on a one-predictor problem where x is the dose of a treatment and y is the binary outcome variable.

[54] I.e., if $Pr(y = 1|x) \geq 0.5$, we conclude $y = 1$; otherwise, $y = 0$.

with a price, and we need to be cautious when we draw conclusions about the linear model, although there are easy conventions for us to follow. On the other hand, it would sound absurd if we dig into the literature and found there had been no *linear model* for binary classification problems. Linear model is the baseline of the data analytics enterprise. It is the starting point of our data analytics adventure. That is how important it is.

Back to the business to modify the linear formalism for a binary classification problem. Now our outcome variable is $Pr(y = 1|x)$, and we realize it still doesn't match with the linear form $\beta_0 + \sum_{i=1}^{p} \beta_i x_i$. What is the essential task here? If we put the puzzle in a context, it may give us some hints. For example, if our goal is to predict the risk of Alzheimer's disease for subjects who are aged 65 years or older, we have known the average risk from recent national statistics is 8.8%. Now if we have a group of individuals who are aged 65 years or older, we could make a risk prediction for them *as a group*, i.e., 8.8%. But this is not the best we could do for each *individual*. We could examine an individual's characteristics such as the gene *APOE*[55] and see if an individual has higher (or lower) risk than the average. Now comes the inspiration: what if we can *rank* the risk of the individuals based on their characteristics, can it help with the final goal that is to predict the outcome variable y?

Now we look closer into the idea of a linear form, and we realize it is more useful in *ranking* the possibilities rather than directly being eligible probabilities.

$$\text{Revised goal: } Pr(y = 1|x) \propto \beta_0 + \sum_{i=1}^{p} \beta_i x_i. \tag{26}$$

In other words, a linear form can make a comparison of two inputs, say, x_i and x_j, and evaluates which one leads to a higher probability of $Pr(y = 1|x)$.

It is fine that we use the linear form to generate numerical values that rank the subjects. We just need one more step to transform those ranks into probabilities. Statisticians have found that the **logistic function** is suitable here for the transformation

$$Pr(y = 1|x) = \frac{1}{1 + e^{-\left(\beta_0 + \sum_{i=1}^{p} \beta_i x_i\right)}}. \tag{27}$$

Figure 27 shows that the *logistic function* indeed provides a better fit of the data than the linear function as shown in Figure 26.

Eq. 27 can be rewritten as Eq. 28

$$\log \frac{Pr(y = 1|x)}{1 - Pr(y = 1|x)} = \beta_0 + \sum_{i=1}^{p} \beta_i x_i. \tag{28}$$

[55] *APOE* polymorphic alleles play a major role in determining the risk of Alzheimer's disease (AD): individuals carrying the $\epsilon 4$ allele are at increased risk of AD compared with those carrying the more common $\epsilon 3$ allele, whereas the $\epsilon 2$ allele decreases risk.

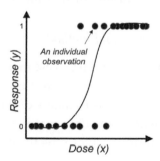

Figure 27: Application of the **logistic function** on binary outcome

This is the so-called **logistic regression** model. The name stems from the transformation of $Pr(y = 1|x)$ used here, i.e., the $\log \frac{Pr(y=1|x)}{1-Pr(y=1|x)}$, which is the logistic transformation that has been widely used in many areas such as physics and signal processing.

Note that we have mentioned that we can predict $y = 1$ if $Pr(y = 1|x) \geq 0.5$, and $y = 0$ if $Pr(y = 1|x) < 0.5$. While 0.5 seems naturally a cut-off value here, it is not necessarily optimal in every application. We could use the techniques discussed in **Chapter 5** such as cross-validation to decide what is the optimal cut-off value in practice.

Visual inspection of data How do we know that our data could be characterized using a logistic function?

We can *discretize* the predictor x in Figure 27 into a few categories, compute the empirical estimate of $Pr(y = 1|x)$ in each category, and create a new data table. This procedure is illustrated in Figure 28.

Suppose that we discretize the data in Figure 27 and obtain the result as shown in Table 6.

	x	y
$Pr(y = 1|x) = 1/3$	0.02	0
	0.03	0
	0.09	1
$Pr(y = 1|x) = 1/2$	0.12	0
	0.14	1
	0.22	1
	...	...

Figure 28: Illustration of the discretization process, e.g., two categories (0.0 − 0.1 and 0.1 − 0.2) of x are shown

Level of x	1	2	3	4	5	6	7	8
$Pr(y = 1|x)$	0.00	0.04	0.09	0.20	0.59	0.89	0.92	0.99

Table 6: Example of a result after discretization

Then, we revise the scale of the y-axis of Figure 27 to be $Pr(y = 1|x)$, and create Figure 29. It could be seen that the empirical curve does fit the form of Eq. 27.

Theory and method

We collect data to estimate the regression parameters of the logistic regression in Eq. 28. Denote the sample size as N. $y \in R^{N \times 1}$ denotes the N measurements of the outcome variable, and $X \in R^{N \times (p+1)}$ denotes the data matrix that includes the N measurements of the p input variables plus the dummy variable for the intercept coefficient β_0. As in a linear regression model, β is the column vector form of the regression parameters.

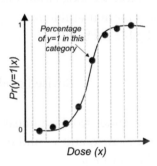

Figure 29: Revised scale of the y-axis of Figure 27, i.e., illustration of Eq. 27

The likelihood function The **likelihood function** evaluates how well a given set of parameters fit the data[56]. The likelihood function has a specific definition, i.e., the conditional probability of the data conditional on the given set of parameters. Here, the dataset is $D = \{X, y\}$, so the likelihood function is defined as $Pr(D|\beta)$. It could be broken down into N components[57]

$$Pr(D|\beta) = \prod_{n=1}^{N} Pr(x_n, y_n|\beta).$$

For data point (x_n, y_n), the conditional probability $Pr(x_n, y_n|\beta)$ is

[56] The *least squares* loss function we derived in **Chapter 2** could also be derived based on the likelihood function of a linear regression model.

[57] Note that, it is assumed that D consists of N independent data points.

$$Pr(x_n, y_n | \beta) = \begin{cases} p(x_n), & if \, y_n = 1 \\ 1 - p(x_n), & if \, y_n = 0. \end{cases} \tag{29}$$

Here, $p(x_n) = Pr(y = 1 | x)$.

A succinct form to represent these two scenarios together is

$$Pr(x_n, y_n | \beta) = p(x_n)^{y_n} [1 - p(x_n)]^{1 - y_n}.$$

Then we can generalize this to all the N data points, and derive the complete likelihood function as

$$Pr(D | \beta) = \prod_{n=1}^{N} p(x_n)^{y_n} [1 - p(x_n)]^{1 - y_n}.$$

It is common to write up its log-likelihood function, defined as $l(\beta) = \log Pr(D | \beta)$, to turn products into sums

$$l(\beta) = \sum_{n=1}^{N} \{ y_n \log p(x_n) + (1 - y_n) \log[1 - p(x_n)] \}.$$

By plugging in the definition of $p(x_n)$, this could be further transformed into

$$l(\beta) = \sum_{n=1}^{N} -\log\left(1 + e^{\beta_0 + \sum_{i=1}^{p} \beta_i x_{ni}}\right) - \sum_{n=1}^{N} y_n (\beta_0 + \sum_{i=1}^{p} \beta_i x_{ni}). \tag{30}$$

Note that, for any probabilistic model[58], we could derive the likelihood function in one way or another, in a similar fashion as we have done for the logistic regression model.

[58] A probabilistic model has a joint distribution for all the random variables concerned in the model. Interested readers can read this comprehensive book: Koller, D. and Friedman, N., *Probabilistic Graphical Models: Principles and Techniques*, The MIT Press, 2009.

Algorithm   Eq. 30 provides the objective function of a maximization problem, i.e., the parameter that maximizes $l(\beta)$ is the best parameter. Theoretically, we could use the First Derivative Test to find the optimal solution. The problem here is that there is no closed-form solution found if we directly apply the First Derivative Test.

Instead, the **Newton-Raphson algorithm** is commonly used to optimize the log-likelihood function of the logistic regression model. It is an iterative algorithm that starts from an **initial solution**, continues to seek updates of the current solution using the following formula

$$\beta^{new} = \beta^{old} - (\frac{\partial^2 l(\beta)}{\partial \beta \partial \beta^T})^{-1} \frac{\partial l(\beta)}{\partial \beta}. \tag{31}$$

Here, $\frac{\partial l(\beta)}{\partial \beta}$ is the **gradient** of the current solution, that points to the **direction** following which we should increment the current solution to improve on the objective function. On the other hand, how far we should go along this direction is decided by the **step size** factor, defined as $\left(\frac{\partial^2 l(\beta)}{\partial \beta \partial \beta^T}\right)^{-1}$. Theoretical results have shown that this formula could converge to the optimal solution. An illustration is given in Figure 30.

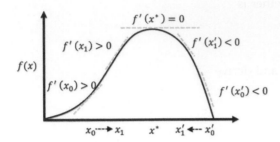

Figure 30: Illustration of the gradient-based optimization algorithms that include the Newton-Raphson algorithm as an example. An algorithm starts from an initial solution (e.g., x_0 and x_0' are two examples of initial solutions in the figure), uses the gradient to find the **direction**, and moves the solution along that direction with the computed **step size**, until it finds the optimal solution x^*.

The Newton-Raphson algorithm presented in Eq. 31 is general. To apply it in a logistic regression model, since we have an explicit form of $l(\beta)$, we can derive the gradient and step size as shown below

$$\frac{\partial l(\beta)}{\partial \beta} = \sum_{n=1}^{N} x_n \left[y_n - p(x_n) \right],$$

$$\frac{\partial^2 l(\beta)}{\partial \beta \partial \beta^T} = -\sum_{n=1}^{N} x_n x_n^T p(x_n) \left[1 - p(x_n) \right].$$

A certain structure can be revealed if we rewrite it in matrix form[59]

[59] $p(x)$ is a $N \times 1$ column vector of $p(x_n)$, and W is a $N \times N$ diagonal matrix with the nth diagonal element as $p(x_n) \left[1 - p(x_n) \right]$.

$$\frac{\partial l(\beta)}{\partial \beta} = X^T \left[y - p(x) \right], \tag{32}$$

$$\frac{\partial^2 l(\beta)}{\beta \beta^T} = -X^T W X. \tag{33}$$

Plugging Eq. 33 into the updating formula as shown in Eq. 31, we can derive a specific formula for logistic regression

$$\beta^{new} = \beta^{old} + (X^T W X)^{-1} X^T \left[y - p(x) \right], \tag{34}$$

$$= (X^T W X)^{-1} X^T W \left(X \beta^{old} + W^{-1} \left[y - p(x) \right] \right), \tag{35}$$

$$= (X^T W X)^{-1} X^T W z. \tag{36}$$

Here, $z = X \beta^{old} + W^{-1}(y - p(x))$.

Putting all these together, a complete flow of the algorithm is shown below

[60] I.e., use random values for β.

1. Initialize β.[60]

2. Compute $p(x_n)$ by its definition: $p(x_n) = \dfrac{1}{1+e^{-(\beta_0 + \sum_{i=1}^{p} \beta_i x_{ni})}}$ for $n = 1, 2, \ldots, N$.

3. Compute the diagonal matrix W, with the nth diagonal element as $p(x_n)[1 - p(x_n)]$ for $n = 1, 2, \ldots, N$.

4. Set z as $= X\beta + W^{-1}[y - p(x)]$.

5. Set $\beta = (X^T W X)^{-1} X^T W z$.

6. If the stopping criteria[61] is met, stop; otherwise, go back to step 2.

Generalized least squares estimator   The estimation formula as shown in Eq. 36 resembles the generalized least squares (GLS) estimator of a regression model, where each data point (x_n, y_n) is associated with a weight w_n. This insight revealed by the Newton-Raphson algorithm suggests a new perspective to look at the logistic regression model. The updating formula shown in Eq. 36 suggests that, in each iteration of parameter updating, we actually solve a weighted regression model as

$$\beta^{new} \leftarrow \arg\min_{\beta} (z - X\beta)^T W (z - X\beta).$$

For this reason, the algorithm we just introduced is also called the **Iteratively Reweighted Least Squares (IRLS)** algorithm. z is referred to as the **adjusted response**.

R Lab

In the AD dataset, the variable `DX_bl` encodes the diagnosis information, i.e., `0` denotes *normal* while `1` denotes *diseased*. We build a logistic regression model using `DX_bl` as the outcome variable.

The 7-Step R Pipeline   **Step 1** is to import data into R.

```
# Step 1 -> Read data into R workstation
# RCurl is the R package to read csv file using a link
library(RCurl)
url <- paste0("https://raw.githubusercontent.com",
              "/analyticsbook/book/main/data/AD.csv")
AD <- read.csv(text=getURL(url))
# str(AD)
```

Step 2 is for data preprocessing.

[61] A common stopping criteria is to evaluate the difference between two consecutive solutions, i.e., if the Euclidean distance between the two vectors, β^{new} and β^{old}, is less than 10^{-4}, then it is considered no difference and the algorithm stops.

```
# Step 2 -> Data preprocessing
# Create your X matrix (predictors) and Y vector (outcome variable)
X <- AD[,2:16]
Y <- AD$DX_bl

# The following code makes sure the variable "DX_bl" is a "factor".
# It denotes "0" as "c0" and "1" as "c1", to highlight the fact
# that "DX_bl" is a factor variable, not a numerical variable.

Y <- paste0("c", Y)
# as.factor is to convert any variable into the
# format as "factor" variable.
Y <- as.factor(Y)

# Then, we integrate everything into a data frame
data <- data.frame(X,Y)
names(data)[16] = c("DX_bl")

set.seed(1) # generate the same random sequence
# Create a training data (half the original data size)
train.ix <- sample(nrow(data),floor( nrow(data)/2) )
data.train <- data[train.ix,]
# Create a testing data (half the original data size)
data.test <- data[-train.ix,]
```

Step 3 is to use the function `glm()` to build a logistic regression model[62].

[62] Type `help(glm)` in R Console to learn more of the function.

```
# Step 3 -> Use glm() function to build a full model
# with all predictors
logit.AD.full <- glm(DX_bl~., data = data.train,
                     family = "binomial")
summary(logit.AD.full)
```

And the result is shown below

```
## Call:
## glm(formula = DX_bl ~ ., family = "binomial", data = data.train)
##
## Deviance Residuals:
##     Min       1Q   Median       3Q      Max
## -2.4250  -0.3645  -0.0704   0.2074   3.1707
##
## Coefficients:
##              Estimate Std. Error z value Pr(>|z|)
## (Intercept)  43.97098    7.83797   5.610 2.02e-08 ***
## AGE          -0.07304    0.03875  -1.885  0.05945 .
## PTGENDER      0.48668    0.46682   1.043  0.29716
## PTEDUCAT     -0.24907    0.08714  -2.858  0.00426 **
## FDG          -3.28887    0.59927  -5.488 4.06e-08 ***
## AV45          2.09311    1.36020   1.539  0.12385
```

```
## HippoNV       -38.03422     6.16738   -6.167 6.96e-10 ***
## e2_1            0.90115     0.85564    1.053  0.29225
## e4_1            0.56917     0.54502    1.044  0.29634
## rs3818361      -0.47249     0.45309   -1.043  0.29703
## rs744373        0.02681     0.44235    0.061  0.95166
## rs11136000     -0.31382     0.46274   -0.678  0.49766
## rs610932        0.55388     0.49832    1.112  0.26635
## rs3851179      -0.18635     0.44872   -0.415  0.67793
## rs3764650      -0.48152     0.54982   -0.876  0.38115
## rs3865444       0.74252     0.45761    1.623  0.10467
## ---
## Signif. codes:  0 '***' 0.001 '**' 0.01 '*' 0.05 '.' 0.1 ' ' 1
##
## (Dispersion parameter for binomial family taken to be 1)
##
##     Null deviance: 349.42  on 257   degrees of freedom
## Residual deviance: 139.58  on 242   degrees of freedom
## AIC: 171.58
##
## Number of Fisher Scoring iterations: 7
```

Step 4 is to use the `step()` function for model selection.

```
# Step 4 -> use step() to automatically delete
# all the insignificant
# variables
# Also means, automatic model selection
logit.AD.reduced <- step(logit.AD.full, direction="both",
                         trace = 0)
summary(logit.AD.reduced)
```

```
## Call:
## glm(formula = DX_bl ~ AGE + PTEDUCAT + FDG + AV45 + HippoNV +
##     rs3865444, family = "binomial", data = data.train)
##
## Deviance Residuals:
##     Min       1Q    Median        3Q       Max
## -2.38957  -0.42407  -0.09268   0.25092   2.73658
##
## Coefficients:
##             Estimate Std. Error z value Pr(>|z|)
## (Intercept) 42.68795    7.07058   6.037 1.57e-09 ***
## AGE         -0.07993    0.03650  -2.190  0.02853 *
## PTEDUCAT    -0.22195    0.08242  -2.693  0.00708 **
## FDG         -3.16994    0.55129  -5.750 8.92e-09 ***
## AV45         2.62670    1.18420   2.218  0.02655 *
## HippoNV    -36.22215    5.53083  -6.549 5.79e-11 ***
## rs3865444    0.71373    0.44290   1.612  0.10707
## ---
## Signif. codes:  0 '***' 0.001 '**' 0.01 '*' 0.05 '.' 0.1 ' ' 1
##
```

```
## (Dispersion parameter for binomial family taken to be 1)
##
##     Null deviance: 349.42  on 257  degrees of freedom
## Residual deviance: 144.62  on 251  degrees of freedom
## AIC: 158.62
##
## Number of Fisher Scoring iterations: 7
```

You may have noticed that some variables included in this model are actually not significant.

Step 4 compares the final model selected by the `step()` function with the full model.

```
# Step 4 continued
anova(logit.AD.reduced,logit.AD.full,test = "LRT")
# The argument, test = "LRT", means that the p-value
# is derived via the Likelihood Ratio Test (LRT).
```

And we can see that the two models are not statistically different, i.e., *p-value* is 0.8305.

Step 5 is to evaluate the overall significance of the final model[63].

```
# Step 5 -> test the significance of the logistic model
# Test residual deviance for lack-of-fit
# (if > 0.10, little-to-no lack-of-fit)
dev.p.val <- 1 - pchisq(logit.AD.reduced$deviance,
                        logit.AD.reduced$df.residual)
```

And it can be seen that the model shows no lack-of-fit as the *p-value* is 1.

```
dev.p.val
```

```
## [1] 1
```

Step 6 is to use your final model for prediction. We can do so using the `predict()` function.

```
# Step 6 -> Predict on test data using your
# logistic regression model
y_hat <- predict(logit.AD.reduced, data.test)
```

Step 7 is to evaluate the prediction performance of the final model.

```
# Step 7 -> Evaluate the prediction performance of
# your logistic regression model
# (1) Three main metrics for classification: Accuracy,
# Sensitivity (1- False Positive),
# Specificity (1 - False Negative)
```

[63] **Step 4** compares two models. **Step 5** tests if a model has a lack-of-fit with data. A model could be better than another, but it is possible that both of them fit the data poorly.

```
y_hat2 <- y_hat
y_hat2[which(y_hat > 0)] = "c1"
# Since y_hat here is the values from the linear equation
# part of the logistic regression model, by default,
# we should use 0 as a cut-off value (only by default,
# not optimal though), i.e., if y_hat < 0, we name it
# as one class, and if y_hat > 0, it is another class.
y_hat2[which(y_hat < 0)] = "c0"

library(caret)
# confusionMatrix() in the package "caret" is a powerful
# function to summarize the prediction performance of a
# classification model, reporting metrics such as Accuracy,
# Sensitivity (1- False Positive),
# Specificity (1 - False Negative), to name a few.
library(e1071)
confusionMatrix(table(y_hat2, data.test$DX_bl))

# (2) ROC curve is another commonly reported metric for
# classification models
library(pROC)
# pROC has the roc() function that is very useful here
plot(roc(data.test$DX_bl, y_hat),
     col="blue", main="ROC Curve")
```

Results are shown below. We haven't discussed the **ROC curve** yet, which will be a main topic in **Chapter 5**. At this moment, remember that a model with a ROC curve that has a larger **Area Under the Curve (AUC)** is a better model. And a model whose ROC curve ties with the diagonal straight line (as shown in Figure 31) is equivalent with random guess.

```
## y_hat2  c0  c1
##     c0 117  29
##     c1  16  97
##
##               Accuracy : 0.8263
##                 95% CI : (0.7745, 0.8704)
##    No Information Rate : 0.5135
##    P-Value [Acc > NIR] : < 2e-16
##
##                  Kappa : 0.6513
##
## Mcnemar's Test P-Value : 0.07364
##
##            Sensitivity : 0.8797
##            Specificity : 0.7698
##         Pos Pred Value : 0.8014
##         Neg Pred Value : 0.8584
##             Prevalence : 0.5135
```

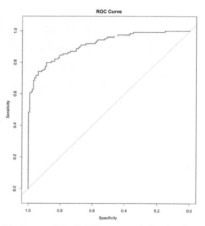

Figure 31: The ROC curve of the final model

```
##         Detection Rate : 0.4517
##   Detection Prevalence : 0.5637
##      Balanced Accuracy : 0.8248
##
##        'Positive' Class : c0
```

Model uncertainty   The 95% confidence interval (CI) of the regression coefficients can be derived, as shown below

```
## coefficients and 95% CI
cbind(coef = coef(logit.AD.reduced), confint(logit.AD.reduced))
```

Results are

```
##                    coef        2.5 %        97.5 %
## (Intercept)   42.68794758   29.9745022   57.88659748
## AGE           -0.07993473   -0.1547680   -0.01059348
## PTEDUCAT      -0.22195425   -0.3905105   -0.06537066
## FDG           -3.16994212   -4.3519800   -2.17636447
## AV45           2.62670085    0.3736259    5.04703489
## HippoNV      -36.22214822  -48.1671093  -26.35100122
## rs3865444      0.71373441   -0.1348687    1.61273264
```

Prediction uncertainty   As in linear regression, we could derive the variance of the estimated regression coefficients $\text{var}(\hat{\beta})$; then, since $\hat{y} = X\hat{\beta}$, we can derive $\text{var}(\hat{y})$[64]. Skipping the technical details, the 95% CI of the predictions are obtained using the R code below

[64] The linearity assumption between x and y enables the explicit characterization of this chain of uncertainty propagation.

```
# Remark: how to obtain the 95% CI of the predictions
y_hat <- predict(logit.AD.reduced, data.test, type = "link",
                 se.fit = TRUE)
# se.fit = TRUE, is to get the standard error in the predictions,
# which is necessary information for us to construct
# the 95% CI of the predictions
data.test$fit    <- y_hat$fit
data.test$se.fit <- y_hat$se.fit
# We can readily convert this information into the 95% CIs
# of the predictions (the way these 95% CIs are
# derived are again, only in approximated sense).
# CI for fitted values
data.test <- within(data.test, {
# added "fitted" to make predictions at appended temp values
fitted    = exp(fit) / (1 + exp(fit))
fit.lower = exp(fit - 1.96 * se.fit) / (1 +
                                 exp(fit - 1.96 * se.fit))
fit.upper = exp(fit + 1.96 * se.fit) / (1 +
                                 exp(fit + 1.96 * se.fit))
})
```

Odds ratio The **odds ratio (OR)** quantifies the strength of the association between two events, *A* and *B*. It is defined as the ratio of the odds of *A* in the presence of *B* and the odds of *A* in the absence of *B*, or equivalently due to symmetry.

- If the *OR* equals 1, *A* and *B* are independent;
- If the *OR* is greater than 1, the presence of one event increases the odds of the other event;
- If the *OR* is less than 1, the presence of one event reduces the odds of the other event.

A regression coefficient of a logistic regression model can be converted into an *odds ratio*, as done in the following codes.

```
## odds ratios and 95% CI
exp(cbind(OR = coef(logit.AD.reduced),
          confint(logit.AD.reduced)))
```

The odds ratios and their 95% CIs are

```
##                        OR         2.5 %        97.5 %
## (Intercept) 3.460510e+18 1.041744e+13 1.379844e+25
## AGE         9.231766e-01 8.566139e-01 9.894624e-01
## PTEDUCAT    8.009520e-01 6.767113e-01 9.367202e-01
## FDG         4.200603e-02 1.288128e-02 1.134532e-01
## AV45        1.382807e+01 1.452993e+00 1.555605e+02
## HippoNV     1.857466e-16 1.205842e-21 3.596711e-12
## rs3865444   2.041601e+00 8.738306e-01 5.016501e+00
```

Exploratory Data Analysis (EDA) EDA essentially conceptualizes the analysis process as a dynamic one, sometimes with a playful tone[65] EDA could start with something simple. For example, we can start with a smaller model rather than throw everything into the analysis.

Let's revisit the data analysis done in the *7-step R pipeline* and examine a simple logistic regression model with only one predictor, `FDG` .

```
# Fit a logistic regression model with FDG
logit.AD.FDG <- glm(DX_bl ~ FDG, data = AD, family = "binomial")
summary(logit.AD.FDG)
```

```
##
## Call:
## glm(formula = DX_bl  FDG, family = "binomial", data = AD)
##
## Deviance Residuals:
##     Min       1Q   Median       3Q      Max
```

[65] And it is probably because of this conceptual framework, EDA happens to use a lot of figures to explore the data. Figures are rich in information, some are not easily generalized into abstract numbers.

```
## -2.4686  -0.8166  -0.2758   0.7679   2.7812
##
## Coefficients:
##            Estimate Std. Error z value Pr(>|z|)
## (Intercept)  18.3300     1.7676   10.37   <2e-16 ***
## FDG          -2.9370     0.2798  -10.50   <2e-16 ***
## ---
## Signif. codes:  0 '***' 0.001 '**' 0.01 '*' 0.05 '.' 0.1 ' ' 1
##
## (Dispersion parameter for binomial family taken to be 1)
##
##     Null deviance: 711.27  on 516  degrees of freedom
## Residual deviance: 499.00  on 515  degrees of freedom
## AIC: 503
##
## Number of Fisher Scoring iterations: 5
```

It can be seen that the predictor `FDG` is significant, as the *p-value* is $< 2e - 16$ that is far less than 0.05. On the other hand, although there is no *R-Squared* in the logistic regression model, we could observe that, out of the total deviance of 711.27, $711.27 - 499.00 = 212.27$ could be explained by `FDG`.

This process could be repeated for every variable in order to have a sense of what are their *marginal* contributions in explaining away the variation in the outcome variable. This practice, which seems dull, is not always associated with an immediate reward. But it is not uncommon in practice, particularly when we have seen in **Chapter 2** that, in regression models, the regression coefficients are interdependent, the regression models are not causal models, and, when you throw variables into the model, they may generate interactions just like chemicals, etc. Looking at your data from every possible angle is useful to conduct data *analytics*.

Back to the simple model that only uses one variable, `FDG`. To understand better how well it predicts the outcome, we can draw figures to visualize the predictions. First, let's get the predictions and their 95% CI values.

```
logit.AD.FDG <- glm(DX_bl ~  FDG, data = data.train,
                    family = "binomial")
y_hat <- predict(logit.AD.FDG, data.test, type = "link",
                se.fit = TRUE)
data.test$fit    <- y_hat$fit
data.test$se.fit <- y_hat$se.fit

# CI for fitted values
data.test <- within(data.test, {
# added "fitted" to make predictions at appended temp values
  fitted     = exp(fit) / (1 + exp(fit))
```

```
    fit.lower = exp(fit - 1.96 * se.fit) / (1 + exp(fit - 1.96 *
                                                    se.fit))
    fit.upper = exp(fit + 1.96 * se.fit) / (1 + exp(fit + 1.96 *
                                                    se.fit))
})
```

We then draw Figure 32 using the following script.

```
# Use Boxplot to evaluate the prediction performance
require(ggplot2)
p <- qplot(factor(data.test$DX_bl), data.test$fit, data = data.test,
    geom=c("boxplot"), fill = factor(data.test$DX_bl)) +
    labs(fill="Dx_bl") +
    theme(text = element_text(size=25))
```

Figure 32 indicates that the model can separate the two classes significantly (while not being good enough). It gives us a global presentation of the prediction. We can draw another figure, Figure 33, to examine more details, i.e., look into the "local" parts of the predictions to see where we can improve.

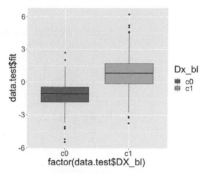

Figure 32: Boxplots of the *predicted probabilities of diseased*, i.e., the $Pr(y = 1|x)$

```
library(ggplot2)
newData <- data.test[order(data.test$FDG),]
newData$DX_bl = as.numeric(newData$DX_bl)
newData$DX_bl[which(newData$DX_bl==1)] = 0
newData$DX_bl[which(newData$DX_bl==2)] = 1
newData$DX_bl = as.numeric(newData$DX_bl)
p <- ggplot(newData, aes(x = FDG, y = DX_bl))
# predicted curve and point-wise 95\% CI
p <- p + geom_ribbon(aes(x = FDG, ymin = fit.lower,
                         ymax = fit.upper), alpha = 0.2)
p <- p + geom_line(aes(x = FDG, y = fitted), colour="red")
# fitted values
p <- p + geom_point(aes(y = fitted), size=2, colour="red")
# observed values
p <- p + geom_point(size = 2)
p <- p + ylab("Probability") + theme(text = element_text(size=18))
p <- p + labs(title =
                "Observed and predicted probability of disease")
print(p)
```

Figure 33 shows that the model captures the relationship between `FDG` with `DX_bl` with a smooth logit curve, and the prediction confidences are fairly small (evidenced by the tight 95% CIs). On the other hand, it is also obvious that the single-predictor model does well on the two ends of the probability range (i.e., close to 0 or 1), but not in the middle range where data points from the two classes could not be clearly separated.

We can add more predictors to enhance its prediction power. To decide on which predictors we should include, we can visualize the

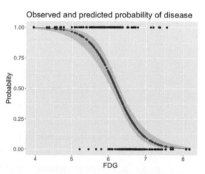

Figure 33: Predicted probabilities (the red curve) with their 95% CIs (the gray area) versus observed outcomes in data (the dots above and below)

relationships between the predictors with the outcome variable. For example, continuous predictors could be presented in **Boxplot** to see if the distribution of the continuous predictor is different across the two classes, i.e., if it is different, it means the predictor could help separate the two classes. The following R codes generate Figure 34.

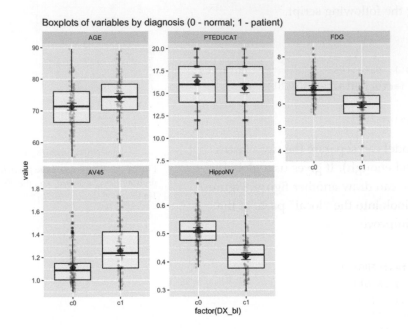

Figure 34: Boxplots of the continuous predictors in the two classes

```
# install.packages("reshape2")
require(reshape2)
data.train$ID <- c(1:dim(data.train)[1])
AD.long <- melt(data.train[,c(1,3,4,5,6,16,17)],
                id.vars = c("ID", "DX_bl"))
# Plot the data using ggplot
require(ggplot2)
p <- ggplot(AD.long, aes(x = factor(DX_bl), y = value))
# boxplot, size=.75 to stand out behind CI
p <- p + geom_boxplot(size = 0.75, alpha = 0.5)
# points for observed data
p <- p + geom_point(position = position_jitter(w = 0.05, h = 0),
                    alpha = 0.1)
# diamond at mean for each group
p <- p + stat_summary(fun = mean, geom = "point", shape = 18,
                    size = 6, alpha = 0.75, colour = "red")
# confidence limits based on normal distribution
p <- p + stat_summary(fun.data = "mean_cl_normal",
                    geom = "errorbar", width = .2, alpha = 0.8)
p <- p + facet_wrap(~ variable, scales = "free_y", ncol = 3)
p <- p + labs(title =
    "Boxplots of variables by diagnosis (0 - normal; 1 - patient)")
print(p)
```

Figure 34 shows that some variables, e.g., `FDG` and `HippoNV`, could separate the two classes significantly. Some variables, such as `AV45` and `AGE`, have less prediction power, but still look promising. Note these observations cautiously, since these figures only show *marginal* relationship among variables[66].

Figure 35 shows the boxplot of the predicted probabilities of diseased made by the final model identified in **Step 4** of the *7-step R pipeline*. This figure is to be compared with Figure 32. It indicates that the final model is much better than the model that only uses the predictor `FDG` alone.

Ranking problem by pairwise comparison

Rationale and formulation

In recent years, we have witnessed a growing interest in estimating the ranks of a list of items. This same problem could be found in a variety of applications, such as the online advertisement of products on Amazon or movie recommendation by Netflix. These problems could be analytically summarized as: given a list of items denoted by $M = \{M_1, M_2, \ldots, M_p\}$, what is the rank of the items (denoted by $\phi = \{\phi_1, \phi_2, \ldots, \phi_p\}$)?[67]

To obtain ranking of items, comparison data (either by domain expert or users) is often collected, e.g., a pair of items in M, let's say, M_i and M_j, will be pushed to the expert/user who conducts the comparison to see if M_i is better than M_j; then, a score, denoted as y_k, will be returned, i.e., a positive y_k indicates that the expert/user supports that M_i is better than M_j, while a negative y_k indicates the opposite. Note that the larger the y_k, the stronger the support.

Denote the expert/user data as y, which is a vector and consists of the set of pairwise comparisons. The question is to estimate the ranking ϕ based on y.

Theory and method

It looks like an unfamiliar problem, but a surprise recognition was made in the paper[68] that the underlying statistical model is a linear regression model. This indicates that we can use the rich array of methods in linear regression framework to solve many problems in ranking.

To see that, first, we need to make explicit the relationship between the parameter to be estimated (ϕ) and the data (y). For the kth comparison that involves items M_i and M_j, we could assume that y_k is

[66] Boxplot is nice but it cannot show synergistic effects among the variables.

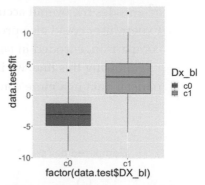

Figure 35: Boxplots of the *predicted probabilities of diseased*, i.e., the $Pr(y = 1|x)$

[67] Here, ϕ is a vector of real values, i.e., the larger the ϕ_i, the higher the rank of M_i.

[68] Osting, B., Brune, C. and Osher, S. *Enhanced statistical rankings via targeted data collection.* Proceedings of the 30th International Conference on Machine Learning (ICML), 2013.

distributed as

$$y_k \sim N\left(\phi_i - \phi_j, \sigma^2/w_k\right). \tag{37}$$

This assumes that if the item M_i is more (or less) important than the item M_j, we will expect to see positive (or negative) values of y_k. σ^2 encodes the overall accuracy level of the expert/user knowledge[69]. Expert/user could also provide their confidence level on a particular comparison, encoded in w_k[70].

Following this line, we illustrate how we could represent the comparison data in a more compact matrix form. This is shown in Figure 36.

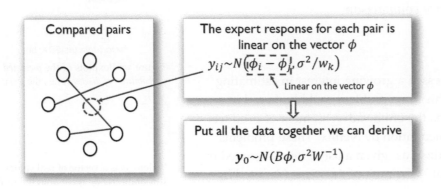

[69] More knowledgeable expert/user will have smaller σ^2.

[70] When this information is lacking, we could simply assume $w_k = 1$ for all the comparison data.

Figure 36: The data structure and its analytic formulation underlying the pairwise comparison. Each node is an item in M, while each arc represents a comparison of two items

The matrix B shown in Figure 36 is defined as

$$B_{kj} = \begin{cases} 1 & \text{if } j = head(k) \\ -1 & \text{if } j = \text{tail}(k) \\ 0 & \text{otherwise} \end{cases}$$

Here, $j = tail(k)$ if the kth comparison is asked in the form as "if M_i is better than M_j" (i.e., denoted as $M_i \rightarrow M_j$); otherwise, $j = head(k)$ for a question asked in the form as $M_j \rightarrow M_i$.

Based on the definition of B, we rewrite Eq. 37 as

$$y_k = \sum_{i=1}^{p} \phi_i B_{ki} + \varepsilon_k, \tag{38}$$

where the distribution of ϵ_k is

$$\epsilon_k \sim N\left(0, \sigma^2/w_k\right). \tag{39}$$

Putting Eq. 38 in matrix form, we can derive that

$$y \sim N\left(B\phi, \sigma^2 W^{-1}\right).$$

where W is the diagonal matrix of elements w_k for $k = 1, 2, \ldots, K$.

[71] Here, the estimation of ϕ is a generalized least squares problem.

Using the framework developed in **Chapter 2**,[71] we could derive the estimator of ϕ as

$$\hat{\phi} = \left(B^T W B\right)^{-1} B^T W y.$$

Statistical process control using decision tree

A fundamental problem in **statistical process control** (**SPC**) is illustrated in Figure 37: given a sequence of observations of a variable that *represents the temporal variability of a process*, is this process *stable*?

SPC is built on a creative application of the statistical distribution theory. A distribution model represents a *stable process*—that is the main premise of SPC—while also allots a calculated proportion to *chance outliers*. An illustration is shown in Figure 38 (left).

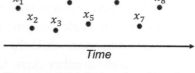

Figure 37: A fundamental problem in statistical process control

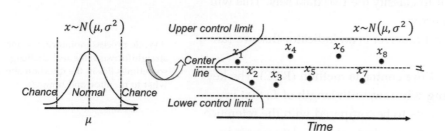

Figure 38: (Left) The use of a distribution model to represent a *stable process*; and (right) the basic idea of a control chart

A further invention of SPC is to convert a distribution model, a static object, into a temporal chart, the so-called **control chart**, as shown in Figure 38 (right). A *control chart* has the upper and lower *control limits*, and a *center line*. It is interesting to note that Figure 38 (left) also provides a graphical illustration of how *hypothesis testing* works, and Figure 38 (right) illustrates the concept of *control chart*. The two build on the same foundation and differ in perspectives: one is horizontal and the other vertical.

A control chart is used to monitor a process. A reference data is collected to draw the control limits and the center line. Then, new data will be continuously collected over time and drawn in the chart, as shown in Figure 39.

Because of this dependency of SPC on distribution models, a considerable amount of interest has been focused on extending it for applications where the data could not be characterized by a distribution. Along this endeavor, how to leverage decision tree models[72] for SPC purposes has been an interesting research problem.

[72] Remember that the decision tree models can deal with complex datasets such as mixed types of variables, as discussed in **Chapter 2**.

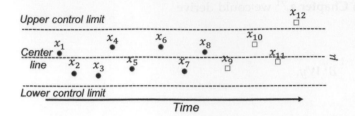

Figure 39: A control chart, built on a reference data (i.e., x_1-x_8), is used to monitor future data (i.e., x_9-x_{12}). An alarm is issued when x_{12} is found to be *out of the control limit*.

Rationale and formulation

One such interesting framework[73] is proposed to cast the process monitoring problem shown in Figure 39 as a classification problem: the *reference data* presumably collected from a stable process represents one class, while the *online data* collected after the reference data represents another class. If the two classes could be significantly separated, an alarm should be issued. Otherwise, if there is no change in the stable process, the two data sets must come from the same distribution, then it will be difficult to classify the two data sets. This will result in a large classification error.

In other words, the classification error is an indicator that we can monitor[74].

Here we introduce the **real-time contrasts** method (**RTC**). The key idea of *RTC* is to have a **sliding window**, with length of L, that includes the most recent data points to be compared with the reference data. We label the *reference data* as one class, and the data points in the *sliding window* as another class. We track the classification error to monitor the process.

We illustrate the *RTC* method through a simple problem. The collected data for monitoring is shown in Table 7. The reference data is $\{1,2\}$.

[73] Deng, H., Runger, G. and Tuv, E., *System monitoring with real-time contrasts*, Journal of Quality Technology, Volume 44, Issue 1, Pages 9-27, 2012.

[74] While process monitoring sounds straightforward, the real challenge sometimes lies in the question about what to monitor, and how.

Data ID	1	2	3	4
Value	2	1	3	3

Table 7: Example of an online dataset with 4 time points

To monitor the process, we use a window size of 2. This means the first monitoring action takes place at the time when the 2nd data point is collected. The reference dataset, $\{1,2\}$, is labeled as class 0, and the two online data points, $\{2,1\}$, are labeled as class 1. As these two datasets are identical, the classification error rate is as large as 0.5. No alarm is issued.

At the next time point, the sliding window now includes data points $\{1,3\}$. A classification rule "*if value* $\leq$ 2, *class* 0; *else, class* 1" would achieve the best classification error rate as 0.25. An alarm probably should be issued.

At the next time point, the sliding window includes data points

$\{3,3\}$. The same classification rule *"if value $\leq$ 2, class 0; else, class 1"* can classify all examples correctly with error rate of 0. An alarm should be issued.

We see that the classification error rate is a **monitoring statistic** to guide the triggering of alerts. It is also useful to use the probability estimates of the data points as the *monitoring statistic*. In other words, the sum of the probability estimates from all data points in the sliding window can be used for monitoring, which is defined as

$$p_t = \frac{\sum_{i=1}^{w} \hat{p}_1(x_i)}{w}.$$

Here, x_i is the ith data point in the sliding window, w is the window size, and $\hat{p}_1(x_i)$ is the probability estimate of x_i belonging to class 1. At each time point in monitoring, we can obtain a p_t. Following the tradition of control chart, we could chart the time series of p_t and observe the patterns to see if alerts should be triggered.

R Lab

We have coded the RTC method into a R function, `Monitoring()`, as shown below, to give an example about how to write self-defined function in R. This function takes two datasets as input: the first is the reference data, `data0`, and the second is the online data points, `data.real.time`. The window size should also be provided in `wsz`. And we use a classification method named random forest[75] to build a classifier. The `Monitoring()` function returns a few monitoring statistics for each online data point, and a score of each variable that represents how likely the variable is responsible for the process change.

[75] More details are in **Chapter 4**.

```
library(dplyr)
library(tidyr)
library(randomForest)
library(ggplot2)

theme_set(theme_gray(base_size = 15) )

# define monitoring function. data0: reference data;
# data.real.time: real-time data; wsz: window size
Monitoring <- function( data0, data.real.time, wsz ){
num.data.points <- nrow(data.real.time)
stat.mat <- NULL
importance.mat <- NULL

for( i in 1:num.data.points ){
# at the start of monitoring, when real-time data size is
# smaller than the window size, combine the real-time
```

```r
# data points and random samples from the reference data
# to form a data set of wsz
if(i<wsz){
  ssfr <- wsz - i
  sample.reference <- data0[sample(nrow(data0),
                                ssfr,replace = TRUE), ]
  current.real.time.data <- rbind(sample.reference,
                          data.real.time[1:i,,drop=FALSE])
}else{
  current.real.time.data <-  data.real.time[(i-wsz+
                                1):i,,drop=FALSE]
}
current.real.time.data$class <- 1
data <- rbind( data0, current.real.time.data )
colnames(data) <- c(paste0("X",1:(ncol(data)-1)),
                "Class")
data$Class <- as.factor(data$Class)

# apply random forests to the data
my.rf <- randomForest(Class ~ .,sampsize=c(wsz,wsz), data=data)

# get importance score
importance.mat <- rbind(importance.mat, t(my.rf$importance))
# get monitoring statistics
ooblist <- my.rf[5]
oobcolumn=matrix(c(ooblist[[1]]),2:3)
ooberrornormal= (oobcolumn[,3])[1]
ooberrorabnormal=(oobcolumn[,3])[2]

temp=my.rf[6]
p1vote <- mean(temp$votes[,2][(nrow(data0)+1):nrow(data)])

this.stat <- c(ooberrornormal,ooberrorabnormal,p1vote)
stat.mat <- rbind(stat.mat, this.stat)
}
result <- list(importance.mat = importance.mat,
            stat.mat = stat.mat)
return(result)
}
```

To demonstrate how to use `Monitoring()` , let's consider a 2-dimensional process with two variables, x_1 and x_2. We simulate the reference data that follow a normal distribution with mean of 0 and standard deviation of 1. The online data come from two distributions: the first 100 data points are sampled from the same distribution as the reference data, while the second 100 data points are sampled from another distribution (i.e., the mean of x_2 changes to 2). We label the reference data with class 0 and the online data with class 1.

```
# data generation
# sizes of reference data, real-time data without change,
# and real-time data with changes
length0 <- 100
length1 <- 100
length2 <- 100

# 2-dimension
dimension <- 2

# reference data
data0 <- rnorm( dimension * length0, mean = 0, sd = 1)
# real-time data with no change
data1 <- rnorm( dimension * length2, mean = 0, sd = 1)
# real-time data different from the reference data in the
# second the variable
data2 <- cbind( V1 = rnorm( 1 * length1, mean = 0, sd = 1),
                V2 = rnorm( 1 * length1, mean = 2, sd = 1) )

# convert to data frame
data0 <- matrix(data0, nrow = length0, byrow = TRUE) %>%
  as.data.frame()
data1 <- matrix(data1, nrow = length2, byrow = TRUE) %>%
  as.data.frame()
data2 <- data2 %>% as.data.frame()

# assign variable names
colnames( data0 ) <- paste0("X",1:ncol(data0))
colnames( data1 ) <- paste0("X",1:ncol(data1))
colnames( data2 ) <- paste0("X",1:ncol(data2))

# assign reference data with class 0 and real-time data with class 1
data0 <- data0 %>% mutate(class = 0)
data1 <- data1 %>% mutate(class = 1)
data2 <- data2 %>% mutate(class = 1)

# real-time data consists of normal data and abnormal data
data.real.time <- rbind(data1,data2)
```

Figure 40 shows the scatterplot of the reference dataset and the first 100 online data points. It can be seen that the two sets of data points are similar.

```
data.plot <- rbind( data0, data1 ) %>% mutate(class = factor(class))
ggplot(data.plot, aes(x=X1, y=X2, shape = class, color=class)) +
  geom_point(size=3)
```

Figure 41 shows the scatterplot of the reference dataset and the second 100 online data points.

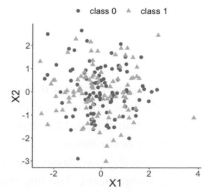

Figure 40: Scatterplot of the reference dataset and the first 100 online data points; both data come from the process under normal condition

```
data.plot <- rbind( data0, data2 ) %>% mutate(class = factor(class))
ggplot(data.plot, aes(x=X1, y=X2, shape = class,
                      color=class)) + geom_point(size=3)
```

Now we apply the *RTC* method. A *window size* of 10 is used. The error rates of the two classes and the probability estimates of the data points over time are shown in Figure 42 drawn by the following R code.

```
wsz <- 10
result <- Monitoring( data0, data.real.time, wsz )
stat.mat <- result$stat.mat
importance.mat <- result$importance.mat

# plot different monitor statistics
stat.mat <- data.frame(stat.mat)
stat.mat$id <- 1:nrow(stat.mat)
colnames(stat.mat) <- c("error0","error1","prob","id")
stat.mat <- stat.mat %>% gather(type, statistics, error0,
                                error1,prob)
ggplot(stat.mat,aes(x=id,y=statistics,color=type)) +
  geom_line(linetype = "dashed") + geom_point() +
  geom_point(size=2)
```

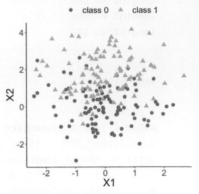

Figure 41: Scatterplot of the reference dataset and the second 100 online data points that come from the process under abnormal condition

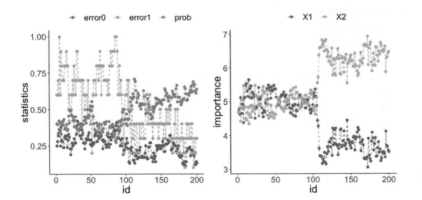

Figure 42: (Left) Chart of the monitoring statistics over time. Three monitoring statistics are shown: *error0* denotes the error rate in Class 0, *error1* denotes the error rate in Class 1, and *prob* denotes the probability estimates of the data points; (right) chart of the importance score of the two variables

We have known that the process shift happened on x_2 after the 100th data point—and that is exactly when a good monitoring statistic should signal the process change. Check Figure 42 (left) and draw your observation.

As the two classes are separated, we could check which variables are significant. The importance scores of the two variables obtained by the random forest model are shown in Figure 42 (right) drawn by the following R code.

```
# plot importance scores for diagnosis
importance.mat <- data.frame(importance.mat)
importance.mat$id <- 1:nrow(importance.mat)
colnames(importance.mat) <- c("X1","X2","id")
importance.mat <- importance.mat %>%
  gather(variable, importance,X1,X2)
ggplot(importance.mat,aes(x=id,y=importance,
    color=variable)) + geom_line(linetype = "dashed") +
    geom_point(size=2)
```

Figure 42 (right) shows that the scores of x_2 significantly increase after the 100th data point. This indicates that x_2 is responsible for the process change, which is true.

Let's consider a 10-dimensional dataset with x_1-x_{10}. We still simulate 100 reference data points of each variable from a normal distribution with mean 0 and variance 1. We use the same distribution to draw the first 100 online data points. Then, we draw the second 100 online data points with two variables, x_9 and x_{10}, whose means changed from 0 to 2.

```
# 10-dimensions, with 2 variables being changed from
# the normal condition
dimension <- 10
wsz <- 5
# reference data
data0 <- rnorm( dimension * length0, mean = 0, sd = 1)
# real-time data with no change
data1 <- rnorm( dimension * length1, mean = 0, sd = 1)
# real-time data different from the reference data in the
# second the variable
data2 <- c( rnorm( (dimension - 2) * length2, mean = 0, sd = 1),
          rnorm( (2) * length2, mean = 20, sd = 1))


# convert to data frame
data0 <- matrix(data0, nrow = length0, byrow = TRUE) %>%
  as.data.frame()
data1 <- matrix(data1, nrow = length1, byrow = TRUE) %>%
  as.data.frame()
data2 <- matrix(data2, ncol = 10, byrow = FALSE) %>%
  as.data.frame()

# assign reference data with class 0 and real-time data
# with class 1
data0 <- data0 %>% mutate(class = 0)
data1 <- data1 %>% mutate(class = 1)
data2 <- data2 %>% mutate(class = 1)

# real-time data consists of normal data and abnormal data
data.real.time <- rbind(data1,data2)
```

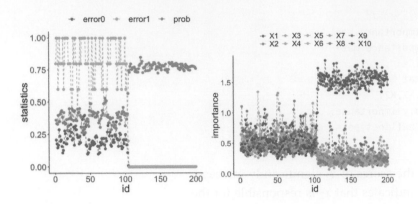

Figure 43: (Left) Chart of the monitoring statistics over time. Three monitoring statistics are shown: *error0* denotes the error rate in Class 0, *error1* denotes the error rate in Class 1, and *prob* denotes the probability estimates of the data points; (right) chart of the importance score of the variables

Figure 43 (left) shows that all the monitoring statistics change after the 101th time point, and the variables' scores in Figure 43 (right) indicate the change is due to x_9 and x_{10}, which is true. The following R codes generated Figure 43 (left).

```
result <- Monitoring( data0, data.real.time, wsz )
stat.mat <- result$stat.mat
importance.mat <- result$importance.mat


# plot different monitor statistics
stat.mat <- data.frame(stat.mat)
stat.mat$id <- 1:nrow(stat.mat)
colnames(stat.mat) <- c("error0","error1","prob","id")
stat.mat <- stat.mat %>% gather(type, statistics, error0,
                                error1,prob)
ggplot(stat.mat,aes(x=id,y=statistics,color=type))+
  geom_line(linetype = "dashed") + geom_point() +
                                geom_point(size=2)
```

The following R codes generated Figure 43 (right).

```
# plot importance scores for diagnosis
importance.mat <- data.frame(importance.mat)
importance.mat$id <- 1:nrow(importance.mat)
# colnames(importance.mat) <- c("X1","X2","id")
importance.mat <- importance.mat %>%
  gather(variable, importance,X1:X10)
importance.mat$variable <- factor( importance.mat$variable,
                          levels = paste0( "X", 1:10))
# levels(importance.mat$variable) <- paste0( "X", 1:10  )
ggplot(importance.mat,aes(x=id,y=importance,color=
          variable)) + geom_line(linetype = "dashed") +
          geom_point(size=2)
```

Remarks

More about the logistic function

Like the linear regression model, Eq. 28 seems like *one* model that explains all the data points[76]. This observation is good, but we may easily overlook its subtle complexity. As shown in Figure 44, the logistic regression model is able to encapsulate a complex relationships between x (the dose) with y (the response to treatment) as one succinct mathematical form. This is remarkable, probably unusual, and unmistakably beautiful.

And the regression coefficients flexibly tune the exact shape of the logistic function for each dataset, as shown in Figure 45.

On the other hand, the logistic function is not the only choice. There are some other options, i.e., Chester Ittner Bliss used the *cumulative normal distribution function* to perform the transformation and called his model the **probit regression** model. There is an interesting discussion of this piece of history in statistics in Chapter 9 of the book[77].

Does the logistic function make sense? — An EDA approach

Figure 29 outlines the main premise of the logistic regression model. It remains unknown whether or not this is a practical assumption. Here, we show how we could evaluate this assumption in a specific dataset. Let's use the AD dataset and pick up the predictor, `HippoNV` , and the outcome variable `DX_bl` .

First, we create a data table like the one shown in Table 6. We discretize the continuous variable `HippoNV` into distinct levels, and compute the prevalence of AD incidences within each level (i.e., the $Pr(y = 1|x)$). The following R code serves this data processing purpose.

```
# Create the frequency table in accordance of categorization
# of HippoNV
temp = quantile(AD$HippoNV,seq(from = 0.05, to = 0.95,
                               by = 0.05))
AD$HippoNV.category <- cut(AD$HippoNV, breaks=c(-Inf,
                                       temp, Inf))
tempData <- data.frame(xtabs(~DX_bl + HippoNV.category,
                        data = AD))
tempData <- tempData[seq(from = 2, to =
                2*length(unique(AD$HippoNV.category)),
                by = 2),]
summary(xtabs(~DX_bl + HippoNV.category, data = AD))
tempData$Total <- colSums(as.matrix(xtabs(~DX_bl +
                HippoNV.category,data = AD)))
```

[76] We have mentioned that a model with this trait is called a *global* model.

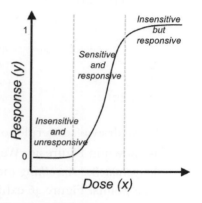

Figure 44: The three regions of the logistic function

[77] Cramer, J.S., *Logit Models from Economics and Other Fields*, Cambridge University Press, 2003.

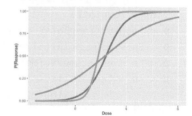

Figure 45: Three examples of the logistic function

```
tempData$p.hat <- 1 - tempData$Freq/tempData$Total
tempData$HippoNV.category = as.numeric(tempData$HippoNV.category)
str(tempData)
```

We use the `str()` function to visualize the data we have converted: 20 levels of `HippoNV` have been created, denoted by the variable `HippoNV.category`; `Total` denotes the total number of subjects within each level; and `p.hat` denotes the proportion of the diseased subjects within each level (i.e., the $Pr(y = 1|x)$).

```
str(tempData)
## 'data.frame':    20 obs. of  5 variables:
## $ DX_bl           : Factor w/ 2 levels "0","1": 2 2 2 2 2 2 ...
## $ HippoNV.category: num  1 2 3 4 5 6 7 8 9 10 ...
## $ Freq            : int  24 25 25 21 22 15 17 17 19 11 ...
## $ Total           : num  26 26 26 26 26 25 26 26 26 34 ...
## $ p.hat           : num  0.0769 0.0385 0.0385 0.1923 0.1538
```

We draw a scatterplot of `HippoNV.category` versus `p.hat`, as shown in Figure 46. We also use the `loess` method, which is a *nonparametric smoothing* method[78], to fit a smooth curve of the scatter data points. Figure 46 exhibits a similar pattern as Figure 29. This provides an empirical justification of the use of the logistic regression model in this dataset.

[78] Related methods will be introduced in **Chapter 9**.

```
# Draw the scatterplot of HippoNV.category
# versus the probability of normal
library(ggplot2)
p <- ggplot(tempData, aes(x = HippoNV.category, y = p.hat))
p <- p + geom_point(size=3)
p <- p + geom_smooth(method = "loess")
p <- p + labs(title ="Empirically observed probability of normal"
              , xlab = "HippoNV")
print(p)
```

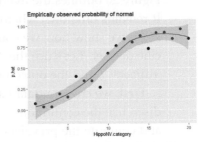

Figure 46: The empirical relationship between `HippoNV` and `DX_bl` takes a shape as the logistic function

Regression vs. tree models

A decision tree model draws a distinct type of **decision boundary**, as illustrated in Figure 47. Think about how a tree is built: at each node, a split is implemented based on *one single variable*, and in Figure 47 the classification boundary is either parallel or perpendicular to one axis.

This implies that, when applying a decision tree to a dataset with linear relationship between predictors and outcome variables, it may not be an optimal choice. In the following example, we simulate a dataset and apply a decision tree and a logistics regression model to the data, respectively. The training data, and the predicted classes for

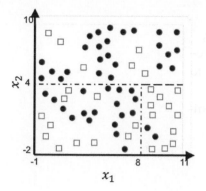

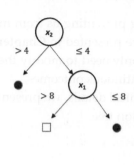

Figure 47: Illustration of a decision tree model for a binary classification problem (i.e., the solid circles and empty squares represent data points from two classes), built on two predictors (i.e., x_1 and x_2); (left) is the scatterplot of the data overlaid with the decision boundary of the decision tree model, which is shown in the (right)

each data point from the logistic regression and decision models are shown in Figures 48, 49 and 50, respectively. It can be seen that the classification boundary from the logistics regression model is linear, while the one from the decision tree is parallel to the axis. Decision tree is not able to capture the linear relationship in the data. The R code for this experiment is shown in below.

```
require(rpart)
ndata <- 2000
X1 <- runif(ndata, min = 0, max = 1)
X2 <- runif(ndata, min = 0, max = 1)
data <- data.frame(X1,X2)
data <- data %>% mutate( X12 = 0.5 * (X1 - X2), Y =
                         ifelse(X12>=0,1,0))
ix <- which( abs(data$X12) <= 0.05)
data$Y[ix] <- ifelse(runif( length(ix)) < 0.5, 0, 1)
data <- data  %>% select(-X12) %>%  mutate(Y =
                         as.factor(as.character(Y)))
ggplot(data,aes(x=X1,y=X2,color=Y))+geom_point()
linear_model <- glm(Y ~  ., family = binomial(link = "logit"),
                         data = data)
tree_model <- rpart( Y ~  ., data = data)
pred_linear <- predict(linear_model, data,type="response")
pred_tree <- predict(tree_model, data,type="prob")[,1]
data_pred <- data %>% mutate(pred_linear_class =
    ifelse(pred_linear <0.5,0,1)) %>%mutate(pred_linear_class =
    as.factor(as.character(pred_linear_class)))%>%
    mutate(pred_tree_class = ifelse( pred_tree <0.5,0,1)) %>%
    mutate( pred_tree_class =
                   as.factor(as.character(pred_tree_class)))
ggplot(data_pred,aes(x=X1,y=X2,color=pred_linear_class))+
  geom_point()
ggplot(data_pred,aes(x=X1,y=X2,color=pred_tree_class))+
  geom_point()
```

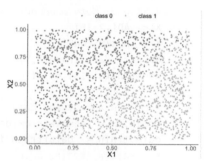

Figure 48: Scatterplot of the generated dataset

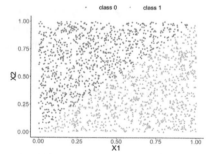

Figure 49: Decision boundary captured by a logistic regression model

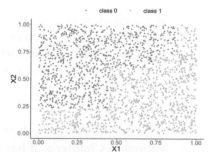

Figure 50: Decision boundary captured by the tree model

Can we use a tree model for regression?

The answer is yes. There is nothing preventing us from modifying the tree-learning process as we have presented in **Chapter 2** for predicting continuous outcome. You only need to modify the IG, i.e., to create a similar counterpart for continuous outcomes.

Without going into further technical details, we present the modified 6-step R pipeline for a regression tree.

```r
# AGE, PTGENDER and PTEDUCAT are used as the
# predictor variables.
# MMSCORE (a numeric value) is the outcome.

# Step 1: read data into R
url <- paste0("https://raw.githubusercontent.com",
              "/analyticsbook/book/main/data/AD.csv")
AD <- read.csv(text=getURL(url))
# Step 2: data preprocessing
X <- AD[,2:16]
Y <- AD$MMSCORE
data <- data.frame(X,Y)
names(data)[16] <- c("MMSCORE")

# Create a training data (half the original data size)
train.ix <- sample(nrow(data),floor( nrow(data)/2) )
data.train <- data[train.ix,]
# Create a testing data (half the original data size)
data.test <- data[-train.ix,]

# Step 3: build the tree
# for regression problems, use method="anova"
tree_reg <- rpart( MMSCORE ~  ., data.train, method="anova")

# Step 4: draw the tree
require(rpart.plot)
prp(tree_reg, nn.cex=1)

# Step 5 -> prune the tree
tree_reg <- prune(tree_reg,cp=0.03)
prp(tree_reg,nn.cex=1)

# Step 6 -> Predict using your tree model
pred.tree <- predict(tree_reg, data.test)
cor(pred.tree, data.test$MMSCORE)
#For regression model, you can use correlation
# to measure how close are your predictions
# with the true outcome values of the data points
```

The learned tree is shown in Figure 51. In the EDA analysis shown in **Chapter 2**, it has been shown that the relationship between `MMSCORE` and `PTEDUCAT` changes substantially according to differ-

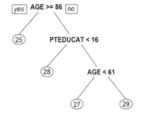

Figure 51: Decision tree to predict `MMSCORE` using `PTEDUCAT` and `AGE`

ent levels of AGE . Here shows the decision tree can also capture the interaction between PTEDUCAT , AGE and MMSCORE .

Exercises

Data analysis

1. Consider the case that, in building linear regression models, there is a concern that some data points may be more important (or more trustable). For these cases, it is not uncommon to assign a weight to each data point. Denote the weight for the ith data point as w_i. An example is shown in Table 8, as the last column, e.g., $w_1 = 1, w_2 = 2, w_5 = 3$.

ID	x_1	x_2	y	w
1	−0.15	−0.48	0.46	1
2	−0.72	−0.54	−0.37	2
3	1.36	−0.91	−0.27	2
4	0.61	1.59	1.35	1
5	−1.11	0.34	−0.11	3

Table 8: Dataset for building a weighted linear regression model

We still want to estimate the regression parameters in the least-squares framework. Follow the process of the derivation of the least-squares estimator as shown in **Chapter 2**, and propose your new estimator of the regression parameters.

2. Follow up the weighted least squares estimator derived in Q1, please calculate the regression parameters of the regression model using the data shown in Table 8.

3. Follow up the dataset in Q1. Use the R pipeline for linear regression on this data (set up the weights in the lm() function). Compare the result from R and the result by your manual calculation.

4. Consider the dataset in Table 5. Use the R pipeline for building a logistic regression model on this data.

5. Consider the model fitted in Q4. Suppose that now there are two new data points as shown in Table 9. Please use the fitted model to predict on these two data points and fill in the table.

ID	x_1	x_2	y
9	0.25	0.18	
10	0.08	1.12	

Table 9: Two test data points

6. Use the dataset PimaIndiansDiabetes2 in the mlbench R package, run the R pipeline for logistic regression on it, and summarize your findings.

7. Follow up on the simulation experiment in Q12 in **Chapter 2**.
   Apply `glm()` on the simulated data to build a logistic regression
   model, and comment on the result.

Chapter 4: Resonance
Bootstrap & Random Forests

Overview

Chapter 4 is about *Resonance*. It is how we work with computers to exploit its remarkable power in conducting iterations and repetitive tasks which human beings find burdensome to do. This capacity of computers enables applications of modern optimization algorithms which underlie many data analytics models. This capacity also makes it realistic to use statistical techniques that don't require analytic tractability.

Decomposing a computational task into repetitive subtasks needs skillful design. The subtasks should be identical in their mathematical forms, so if they differ from each other they only differ in their specific *parametric* configurations. A subtask should also be easy to solve, and sometimes there is even closed-form solution. These subtasks may or may not be carried out in a sequential manner, but as a whole, they *move things forward*—solving a problem. Not all repetitions move things forward, e.g., two types of repetitions are shown in Figures 52 and 53.

Comparing with the *divide-and-conquer* that is more of a spatial nature, here, repetition is more of a temporal nature, and effective designs of repetitions need the power of *resonance* between the repetitions. If they are carried out in a sequential manner, they do not depend on each other logically like in a deduction sequence; and if one subtask is only solved suboptimally, as a whole they move forward towards an optimal solution.

A particular invention that has played a critical role in many data analytic applications is the **Bootstrap**. Building on the idea of Bootstrap, Random Forest was invented in 2001,[79] after that came countless **Ensemble Learning** methods[80].

Figure 52: Iterations from a pendulum

Figure 53: Iterations that form a spiral

[79] Breiman, L., *Random Forests*. Machine Learning, Volume 45, Issue 1, Pages 5-32, 2001.
[80] See **Chapter 7**.

How bootstrap works

Rationale and formulation

There are multiple perspectives to look at Bootstrap. One perspective that has been well studied in the seminar book[81] is to treat Bootstrap as a simulation of the *sampling process*. As we know, sampling refers to the idea that we could draw samples again and again from the same population. Many statistical techniques make sense only when we put them in the framework of sampling[82]. For example, in hypothesis testing, when the Type 1 Error (a.k.a., the α) is set to be 0.05, it means that if we are able to repeat for multiple times the data collection, computation of statistics, and hypothesis testing, on average we will reject the null hypothesis 5% of the times even when the null hypothesis is true.

For many statistical models, the analytic *tractability* of the sampling process lays the foundation to study their behavior and performance. When there were no computers, analytical tractability had been (and still is) one main factor that determines the "fate" of a statistical model—i.e., if we haven't found an analytical tractable formulation to study the model considering its sampling process, it is hard to convince statisticians that the model is valid[83]. Nonetheless, there have been good statistical models that have no rigorous mathematical formulations, yet they are effective in applications. It may take years for us to find a mathematical framework to establish these models as rigorous approaches.

As a remedy, Bootstrap exploits the power of computers to simulate the sampling process. Figure 54 provides a simple illustration of its idea. The input of a Bootstrap algorithm is a dataset, which provides a *representation* of the underlying population. As the dataset is considered to be an representational *equivalent* of the underlying population, *sampling from the underlying population* could be approximated by *resampling from the dataset*. As shown in Figure 54, we could generate any number of Bootstrapped datasets by randomly drawing samples (with or without replacement, both have been useful) from the dataset. The idea is simple and effective.

The conceptual power of Bootstrap   A good model has a concrete procedure of operations. That is *what it does*. There is also a conceptual-level perspective that concerns a model regarding *what it is*.

[81] Efron, B. and Tibshirani, R.J., *An Introduction to the Bootstrap.* Chapman & Hall/CRC, 1993.

[82] When statistics gained its scientific foundation and a modern appeal, there was also the rise of *mass production* that introduced human beings into a mechanical world populated with endless repetitive movements of machines and processes, as dramatized in Charlie Chaplin's movies.

[83] E.g., without the possibility (even the possibility is probably only *theoretical*) of infinite repetition of the same process, the concept α in hypothesis testing will lose its ground of being something tangible.

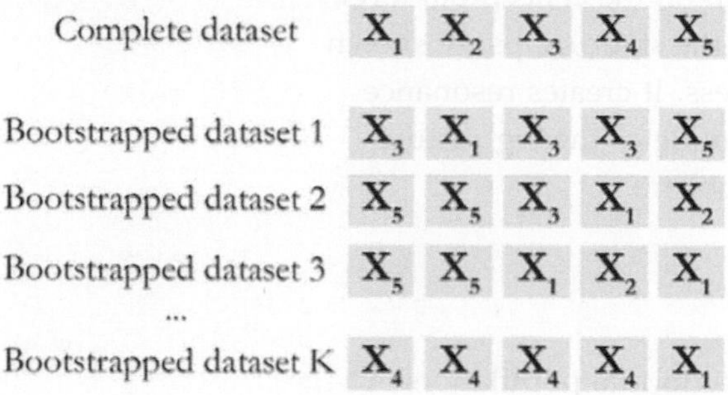

Complete dataset X_1 X_2 X_3 X_4 X_5
Bootstrapped dataset 1 X_3 X_1 X_3 X_3 X_5
Bootstrapped dataset 2 X_5 X_5 X_3 X_1 X_2
Bootstrapped dataset 3 X_5 X_5 X_1 X_2 X_1
...
Bootstrapped dataset K X_4 X_4 X_4 X_4 X_1

Figure 54: A demonstration of the Bootstrap process

A sampling process is conventionally—or conveniently—conceived as a static snapshot, as shown in Figure 55. The data points are pictured as silhouettes of apples, to highlight the psychological tendency we all share that we too often focus on the samples (the fruit) rather than the sampling process (the tree). This view is not entirely wrong, but it is a reduced view and it is easy for this fact to slip below the level of consciousness. We often take the apples as an absolute fact and forget that they are only historic coincidence: they are a *representation* of the apple tree (here, corresponds to the concept *population*); they themselves are not the apple tree.

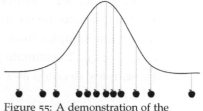

Figure 55: A demonstration of the sampling process

...
POPULATION
Sampling process
SAMPLE
A dynamic view of Bootstrap
POPULATION
Sampling process
SAMPLE
...

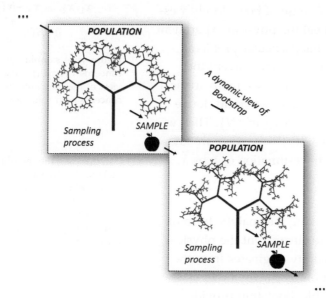

Figure 56: A demonstration of the dynamic view of Bootstrap. The tree is drawn using `http://mwskirpan.com/FractalTree/`.

The Bootstrap starts from what we have forgotten. Not to study an apple as an apple, it studies the process of how an apple is created, the process of apple-*ing*. Bearing this objective in mind, an apple is no longer empirical, now it is a *general* apple. It bears the genetics of apple that is shared by all apples and conceives the possibility of an apple tree. This is the power of the Bootstrap on the conceptual level. As illustrated in Figure 56, it now animates the static snapshot shown in Figure 55 and recovers the dynamic process. It creates resonance among the snapshots, and the resonance completes the big picture and enlarges our conceptual view of the problem.

Theory and method

Let's consider the estimator of the mean of a normal population. A random variable X follows a normal distribution, i.e., $X \sim N\left(\mu, \sigma^2\right)$. For simplicity, let's assume that we have known the variance σ^2. We want to estimate the mean μ. What we need to do is to randomly draw a few samples from the distribution. Denote these samples as $x_1, x_2, \ldots, x_N$. To estimate μ, it seems natural to use the average of the samples, denoted as $\bar{x} = \frac{1}{N} \sum_{n=1}^{N} x_i$. We use $\bar{x}$ as an estimator of μ, i.e., $\hat{\mu} = \bar{x}$.

A question arises, how good is $\bar{x}$ to be an estimator of μ?

Obviously, if $\bar{x}$ is numerically close to μ, it is a good estimator. The problem is, for a particular dataset, we don't know what is μ. And even if we know μ, we need to have criteria to tell us how close is close enough. On top of all these considerations, common sense tells us that $\bar{x}$ itself is a random variable that is subject to uncertainty. To evaluate this uncertainty and get a general sense of how closely $\bar{x}$ estimates μ, a brute-force approach is to repeat the physical experiment many times. But this is not necessary. In this particular problem, where X follows a normal distribution, we could circumvent the physical burden (i.e., of repeating the experiments) via mathematical derivation. Since X follows a normal distribution, we could derive that $\bar{x}$ is another normal distribution, i.e., $\bar{x} \sim N\left(\mu, \sigma^2/N\right)$. Then we know that $\bar{x}$ is an unbiased estimator, since $E(\bar{x}) = \mu$. Also, we know that the larger the sample size, the better the estimation of μ by $\bar{x}$, since the variance of the estimator is σ^2/N.[84]

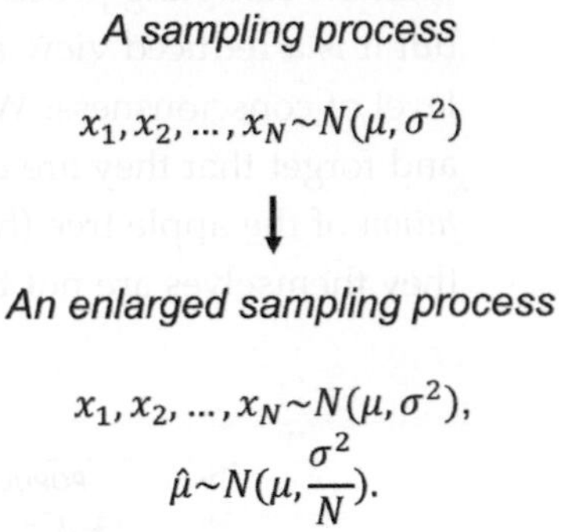

When Bootstrap is needed   Apparently, knowing the analytic form of $\bar{x}$ is the key in the example shown above to circumvent the physical need of sampling. And the analytic tractability originates from the condition that X follows normal distribution. For many other estimators that we found analytically intractable, Bootstrap provides a computational remedy that enables us to investigate their properties, because it *computationally* mimics the physical sampling process.

For example, while the distribution of X is unknown, we could follow the Bootstrap scheme illustrated in Figure 58 to evaluate the sampling distribution of $\bar{x}$. For any bootstrapped dataset, we can calculate the $\bar{x}$ and obtain a "sample" of $\bar{x}$, denoted as $\bar{x}_i$. Figure 58

A sampling process

$$x_1, x_2, \ldots, x_N \sim N(\mu, \sigma^2)$$

$$\downarrow$$

An enlarged sampling process

$$x_1, x_2, \ldots, x_N \sim N(\mu, \sigma^2),$$
$$\hat{\mu} \sim N(\mu, \frac{\sigma^2}{N}).$$

Figure 57: A sampling process that concerns x (upper) and an enlarged view of the sampling process that concerns both x and $\hat{\mu}$ (bottom)

[84] In this case, the sampling process includes (1) drawing samples from the distribution; and (2) estimating μ using $\bar{x}$. Illustration is shown in Figure 57. The *sampling process* is a flexible concept, depending on what variables are under study.

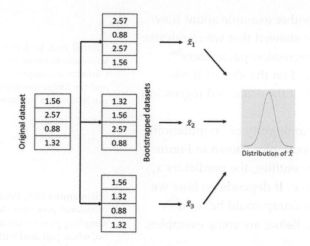

Figure 58: The (nonparametric) Bootstrap scheme to computationally evaluate the sampling distribution of $\bar{x}$

only illustrates three cases, while in practice usually thousands of bootstrapped datasets are drawn. After we have the "samples" of $\bar{x}$, we can draw the distribution of $\bar{x}$ and present it as shown in Figure 58. Although we don't know the distribution's analytic form, we have its numerical representation stored in a computer.

The Bootstrap scheme illustrated in Figure 58 is called *nonparametric* Bootstrap, since no *parametric* model is used to mediate the process[85]. This is not the only way through which we can conduct Bootstrap. For example, a parametric Bootstrap scheme is illustrated in Figure 59 to perform the same task, i.e., to study the sampling distribution of $\bar{x}$. The difference between the nonparametric Bootstrap scheme in Figure 58 and the parametric Bootstrap scheme in Figure 59 is that, when generating new samples, the nonparametric Bootstrap uses the original dataset as the representation of the underlying population, while the parametric Bootstrap uses a fitted distribution model.

[85] Put simply, a parametric model is a model with an explicit mathematical form that is calibrated by parameters.

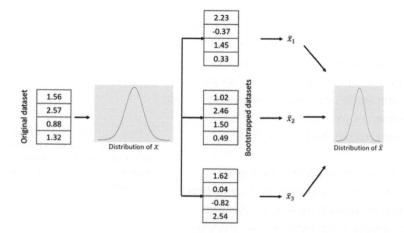

Figure 59: The (parametric) Bootstrap scheme to computationally evaluate the sampling distribution of $\bar{x}$

Bootstrap for regression model   We show another example about how Bootstrap could be used. In **Chapter 2**, we showed that we can derive the explicit distribution of the estimated regression parameters[86]. Here, we introduce another approach, based on the idea of Bootstrap, to *compute* an empirical distribution of the estimated regression parameters.

The first challenge we encounter is the ambiguity of "population" here. Unlike in the parameter estimation examples shown in Figure 58, here, a regression model involves three entities, the predictors x, the outcome variable y, and the error term ϵ. It depends on how we define the "population" to decide how Bootstrap could be used[87].

A variety of options could be obtained. Below are some examples.

Option 1.   We could simply resample the data points (i.e., the (x_n, y_n) pairs) following the nonparametric Bootstrap scheme shown in Figure 58. For each sampled dataset, we fit a regression model and obtain the fitted regression parameters.

Option 2.   We could fix the x, and only sample for y.[88] To sample y, we draw samples using a fitted conditional distribution model $P(y|x)$.[89] Then, for each sampled dataset, we fit a regression model and obtain the fitted parameters.

Option 3.   We could simulate new samples of x using the nonparametric Bootstrap method on the samples of x only. Then, for the new samples of x, we draw samples of y using the fitted conditional distribution model $P(y|x)$. This is a combination of the nonparametric and parametric Bootstrap methods. Then, for each sampled dataset, we can fit a regression model and obtain the fitted regression parameters.

Via either option, we could obtain an empirical distribution of the estimated regression parameter and compute a curve like the one shown in Figure 58. For instance, suppose we repeat this process $10,000$ times. We can obtain $10,000$ sets of estimated regression parameters, then we can use these samples to evaluate the sampling distribution of the regression parameters. We can also see if the parameters are significantly different from 0, and derive their 95% confidence intervals.

The three options above are just some examples[90]. As a more complicated model than simple parametric estimation in distribution fitting, how to conduct Bootstrap on regression models (and other complex models such as time series models or decision tree models) is a challenging problem.

[86] Recall that, to derive this distribution, a few assumptions are needed: the Gaussian assumption of the error term, and the linear assumptions between predictors and outcome variable.

[87] Remember that, Bootstrap is a computational procedure to mimic the sampling process from a *population*. So, when you deal with a problem, you need to determine first what is the population.

[88] In this way we implicitly assume that the uncertainty of the dataset mainly comes from y.

[89] We could use kernel density estimation approaches, which are not introduced in this book. Interested readers could explore the R package `kdensity` and read the monograph by Silverman, B.W., *Density Estimation for Statistics and Data Analysis*, Chapman & Hall/CRC, 1986. A related method, called *Kernel regression* model, can be found in **Chapter 9**.

[90] Options 1 and 3 both define the population as the joint distribution of x and y, but differ in the ways to draw samples. Option 2 defines the population as $P(y|x)$ only.

R Lab

4-Step R Pipeline   **Step 1** is to load the dataset into the R workspace.

```
# Step 1 -> Read data into R workstation
# RCurl is the R package to read csv file using a link
library(RCurl)
url <- paste0("https://raw.githubusercontent.com",
              "/analyticsbook/book/main/data/AD.csv")
AD <- read.csv(text=getURL(url))
```

Step 2 is to implement Bootstrap on a model. Here, let's implement Bootstrap on the parameter estimation problem for normal distribution fitting. We obtain results using both the analytic approach and the Bootstrapped approach, so we could evaluate how well the Bootstrap works.

Specifically, let's pick up the variable `HippoNV`, and estimate its mean for the normal subjects[91]. Assuming that the variable `HippoNV` is distributed as a normal distribution, we could use the `fitdistr()` function from the R package `MASS` to estimate the mean and standard derivation, as shown below.

[91] Recall that we have both *normal* and *diseased* populations in our dataset.

```
# Step 2 -> Decide on the statistical operation
# that you want to "Bootstrap" with
require(MASS)
fit <- fitdistr(AD$HippoNV, densfun="normal")
# fitdistr() is a function from the package "MASS".
# It can fit a range of distributions, e.g., by using the argument,
# densfun="normal", we fit a normal distribution.
```

The `fitdistr()` function returns the estimated parameters together with their standard derivation[92], and the 95% CI of the estimated mean.

[92] Here, the standard derivation of the estimated parameters is derived based on the assumption of normality of `HippoNV`. This is a theoretical result, in contrast with the computational result by Bootstrap shown later.

```
fit
##       mean           sd
##   0.471662891   0.076455789
##  (0.003362522) (0.002377662)
lower.bound = fit$estimate[1] - 1.96 * fit$sd[2]
upper.bound = fit$estimate[1] + 1.96 * fit$sd[2]
##   lower.bound   upper.bound
##    0.4670027     0.4763231
```

Step 3 implements the nonparametric Bootstrap scheme[93], as an alternative approach, to obtain the 95% CI of the estimated mean.

[93] The one shown in Figure 58.

```
# Step 3 -> draw R bootstrap replicates to
# conduct the selected statistical operation
R <- 10000
```

```r
# Initialize the vector to store the bootstrapped estimates
bs_mean <- rep(NA, R)
# draw R bootstrap resamples and obtain the estimates
for (i in 1:R) {
  resam1 <- sample(AD$HippoNV, length(AD$HippoNV),
                   replace = TRUE)
  # resam1 is a bootstrapped dataset.
  fit <- fitdistr(resam1 , densfun="normal")
  # store the bootstrapped estimates of the mean
  bs_mean[i] <- fit$estimate[1]
}
```

Here, 10,000 replications are simulated by the Bootstrap method. The `bs_mean` is a vector of 10,000 elements to record all the estimated mean parameter in these replications. These 10,000 estimated parameters could be taken as a set of samples.

Step 4 is to summarize the Bootstrapped samples, i.e., to compute the 95% CI of the estimated mean, as shown below.

```r
# Step 4 -> Summarize the results and derive the
# bootstrap confidence interval (CI) of the parameter
# sort the mean estimates to obtain quantiles needed
# to construct the CIs
bs_mean.sorted <- sort(bs_mean)
# 0.025th and 0.975th quantile gives equal-tail bootstrap CI
CI.bs <- c(bs_mean.sorted[round(0.025*R)],
                        bs_mean.sorted[round(0.975*R+1)])
CI.bs
##    lower.bound    upper.bound
##     0.4656406      0.4778276
```

It is seen that the 95% CI by Bootstrap is close to the 95% CI in the theoretical result. This shows the validity and efficacy of the Bootstrap method to evaluate the uncertainty of a statistical operation[94].

Beyond the 4-step Pipeline While the estimation of the mean of `HippoNV` is a relatively simple operation, in what follows, we consider a more complex statistical operation, the comparison of the mean parameters of `HippoNV` across the two classes, *normal* and *diseased*.

To do so, the following R code creates a temporary dataset for this purpose.

```r
tempData <- data.frame(AD$HippoNV,AD$DX_bl)
names(tempData) = c("HippoNV","DX_bl")
tempData$DX_bl[which(tempData$DX_bl==0)] <- c("Normal")
tempData$DX_bl[which(tempData$DX_bl==1)] <- c("Diseased")
```

We then use `ggplot()` to visualize the two distributions by comparing their histograms.

[94] Bootstrap is as good as the theoretical method, but it doesn't require to know the variable's distribution. The cost it pays for this robustness is its computational overhead. In this example, the computation is light. For some other cases, the computation could be burdensome, i.e., when the dataset becomes big, or the statistical model itself has been computationally demanding.

```
p <- ggplot(tempData,aes(x = HippoNV, colour=DX_bl))
p <- p +  geom_histogram(aes(y = ..count.., fill=DX_bl),
                 alpha=0.5,position="identity")
print(p)
```

The result is shown in Figure 60. It could be seen that the two distributions differ from each other. To have a formal evaluation of this impression, the following R code shows how the nonparametric Bootstrap method can be implemented here.

```
# draw R bootstrap replicates
R <- 10000
# init location for bootstrap samples
bs0_mean <- rep(NA, R)
bs1_mean <- rep(NA, R)
# draw R bootstrap resamples and obtain the estimates
for (i in 1:R) {
resam0 <- sample(tempData$HippoNV[which(tempData$DX_bl==
  "Normal")],length(tempData$HippoNV[which(tempData$DX_bl==
                    "Normal")]),replace = TRUE)
fit0 <- fitdistr(resam0 , densfun="normal")
bs0_mean[i] <- fit0$estimate[1]
resam1 <- sample(tempData$HippoNV[which(tempData$DX_bl==
            "Diseased")],
        length(tempData$HippoNV[which(tempData$DX_bl==
            "Diseased")]),replace = TRUE)
fit1 <- fitdistr(resam1 , densfun="normal")
bs1_mean[i] <- fit1$estimate[1]
}

bs_meanDiff <- bs0_mean - bs1_mean

# sort the mean estimates to obtain bootstrap CI
bs_meanDiff.sorted <- sort(bs_meanDiff)
# 0.025th and 0.975th quantile gives equal-tail bootstrap CI
CI.bs <- c(bs_meanDiff.sorted[round(0.025*R)],
        bs_meanDiff.sorted[round(0.975*R+1)])
CI.bs
```

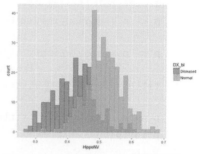

Figure 60: Histograms of `HippoNV` in the *normal* and *diseased* groups

The 95% CI of the difference of the two mean parameters is

```
CI.bs
## [1] 0.08066058 0.10230428
```

The following R code draws a histogram of the `bs_meanDiff` to give us visual information about the Bootstrapped estimation of the mean difference, which is shown in Figure 61. The difference is statistically significant.

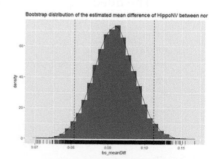

Figure 61: Histogram of the estimated mean difference of `HippoNV` in the two groups by Bootstrap with 10,000 replications

```
## Plot the bootstrap distribution with CI
# First put data in data.frame for ggplot()
dat.bs_meanDiff <- data.frame(bs_meanDiff)

library(ggplot2)
p <- ggplot(dat.bs_meanDiff, aes(x = bs_meanDiff))
p <- p + geom_histogram(aes(y=..density..))
p <- p + geom_density(alpha=0.1, fill="white")
p <- p + geom_rug()
# vertical line at CI
p <- p + geom_vline(xintercept=CI.bs[1], colour="blue",
                    linetype="longdash")
p <- p + geom_vline(xintercept=CI.bs[2], colour="blue",
                    linetype="longdash")
title = "Bootstrap distribution of the estimated mean
         difference of HippoNV between normal and diseased"
p <- p + labs(title =title)
print(p)
```

We can also apply Bootstrap on linear regression. In **Chapter 2** we have fitted a regression model of `MMSCORE` and presented the analytically derived standard derivation of the estimated regression parameters[95]. Here, we show that we could use Bootstrap to compute the 95% CI of the regression parameters as well.

First, we fit a linear regression model.

[95] I.e., as shown in Eq. 20.

```
# Fit a regression model first, for comparison
tempData <- data.frame(AD$MMSCORE,AD$AGE, AD$PTGENDER, AD$PTEDUCAT)
names(tempData) <- c("MMSCORE","AGE","PTGENDER","PTEDUCAT")
lm.AD <- lm(MMSCORE ~ AGE + PTGENDER + PTEDUCAT, data = tempData)
sum.lm.AD <- summary(lm.AD)
# Age is not significant according to the p-value
std.lm <- sum.lm.AD$coefficients[ , 2]
lm.AD$coefficients[2] - 1.96 * std.lm[2]
lm.AD$coefficients[2] + 1.96 * std.lm[2]
```

The fitted regression model is

```
##
## Call:
## lm(formula = MMSCORE   AGE + PTGENDER + PTEDUCAT,
##               data = tempData)
##
## Residuals:
##    Min     1Q  Median     3Q    Max
## -8.4290 -0.9766  0.5796  1.4252  3.4539
##
## Coefficients:
##             Estimate Std. Error t value Pr(>|t|)
## (Intercept) 27.70377    1.11131  24.929  < 2e-16 ***
```

```
## AGE          -0.02453    0.01282   -1.913   0.0563 .
## PTGENDER     -0.43356    0.18740   -2.314   0.0211 *
## PTEDUCAT      0.17120    0.03432    4.988 8.35e-07 ***
## ---
## Signif. codes:  0 '***' 0.001 '**' 0.01 '*' 0.05
## '.' 0.1 ' ' 1
##
## Residual standard error: 2.062 on 513 degrees of freedom
## Multiple R-squared:  0.0612, Adjusted R-squared:  0.05571
## F-statistic: 11.15 on 3 and 513 DF,  p-value: 4.245e-07
##     Lower bound   Upper bound
## [1] -0.04966834   0.000600785
```

Then, we follow Option 1 to conduct Bootstrap for the linear regression model.

```
# draw R bootstrap replicates
R <- 10000
# init location for bootstrap samples
bs_lm.AD_demo <- matrix(NA, nrow = R, ncol =
                        length(lm.AD_demo$coefficients))
# draw R bootstrap resamples and obtain the estimates
for (i in 1:R) {
  resam_ID <- sample(c(1:dim(tempData)[1]), dim(tempData)[1],
                 replace = TRUE)
resam_Data <- tempData[resam_ID,]
  bs.lm.AD_demo <- lm(MMSCORE ~  AGE + PTGENDER + PTEDUCAT,
                 data = resam_Data)
bs_lm.AD_demo[i,] <- bs.lm.AD_demo$coefficients
}
```

The `bs_lm.AD_demo` records the estimated regression parameters in the 10,000 replications. The following R code shows the 95% CI of `AGE` by Bootstrap.

```
bs.AGE <- bs_lm.AD_demo[,2]
# sort the mean estimates of AGE to obtain bootstrap CI
bs.AGE.sorted <- sort(bs.AGE)

# 0.025th and 0.975th quantile gives equal-tail
# bootstrap CI
CI.bs <- c(bs.AGE.sorted[round(0.025*R)],
      bs.AGE.sorted[round(0.975*R+1)])
CI.bs
```

It is clear that the 95% CI of `AGE` includes 0 in the range. This is consistent with the result by t-test that shows the variable `AGE` is insignificant (i.e., p-value= 0.0563). We can also see that the 95% CI by Bootstrap is close to the 95% CI by theoretical result.

```
CI.bs
##      Lower bound    Upper bound
## [1] -0.053940482   0.005090523
```

The following R codes draw a histogram of the Bootstrapped estimation of the regression parameter of `AGE` to give us a visual examination about the Bootstrapped estimation, which is shown in Figure 62.

```
## Plot the bootstrap distribution with CI
# First put data in data.frame for ggplot()
dat.bs.AGE <- data.frame(bs.AGE.sorted)

library(ggplot2)
p <- ggplot(dat.bs.AGE, aes(x = bs.AGE))
p <- p + geom_histogram(aes(y=..density..))
p <- p + geom_density(alpha=0.1, fill="white")
p <- p + geom_rug()
# vertical line at CI
p <- p + geom_vline(xintercept=CI.bs[1], colour="blue",
                    linetype="longdash")
p <- p + geom_vline(xintercept=CI.bs[2], colour="blue",
                    linetype="longdash")
title <- "Bootstrap distribution of the estimated
                    regression parameter of AGE"
p <- p + labs(title = title)
print(p)
```

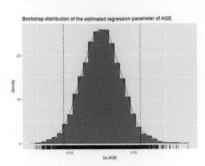

Figure 62: Histogram of the estimated regression parameter of `AGE` by Bootstrap with 10,000 replications

We can also see the 95% CI of `PTEDUCAT` as shown below, which is between 0.1021189 and 0.2429209. This is also close to the the 95% CI by theoretical result. Also, t-test also shows the variable `PTEDUCAT` is significant (i.e., p-value is $8.35e - 07$).

```
bs.PTEDUCAT <- bs_lm.AD_demo[,4]
# sort the mean estimates of PTEDUCAT to obtain
# bootstrap CI
bs.PTEDUCAT.sorted <- sort(bs.PTEDUCAT)

# 0.025th and 0.975th quantile gives equal-tail
# bootstrap CI
CI.bs <- c(bs.PTEDUCAT.sorted[round(0.025*R)],
           bs.PTEDUCAT.sorted[round(0.975*R+1)])
CI.bs
CI.bs
## [1] 0.1021189 0.2429209
```

The following R code draws a histogram (i.e., Figure 63) of the Bootstrapped estimation of the regression parameter of `PTEDUCAT`.

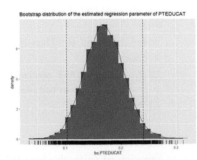

Figure 63: Histogram of the estimated regression parameter of `PTEDUCAT` by Bootstrap with 10,000 replications

```
## Plot the bootstrap distribution with CI
# First put data in data.frame for ggplot()
dat.bs.PTEDUCAT <- data.frame(bs.PTEDUCAT.sorted)

library(ggplot2)
p <- ggplot(dat.bs.PTEDUCAT, aes(x = bs.PTEDUCAT))
p <- p + geom_histogram(aes(y=..density..))
p <- p + geom_density(alpha=0.1, fill="white")
p <- p + geom_rug()
# vertical line at CI
p <- p + geom_vline(xintercept=CI.bs[1], colour="blue",
                    linetype="longdash")
p <- p + geom_vline(xintercept=CI.bs[2], colour="blue",
                    linetype="longdash")
title <- "Bootstrap distribution of the estimated regression
                             parameter of PTEDUCAT"
p <- p + labs(title = title )
print(p)
```

Random forests

Rationale and formulation

Randomness has a productive dimension, depending on how you use it. For example, a **random forest (RF)** model consists of multiple tree models that are generated by a creative use of two types of randomness. One, each tree is built on a *randomly selected* set of samples by applying Bootstrap on the original dataset. Two, in building each tree, a *randomly selected* subset of features is used to choose the best split. Figure 64 shows this scheme of random forest.

Random forest has gained superior performances in many applications, and it (together with its variants) has been a winning approach in some data competitions over the past years. While it is not necessary that an aggregation of many models would lead to better performance than its constituting parts, random forest works because of a number of reasons. Here we use an example to show when the random forest, as a sum, is better than its parts (i.e., the decision trees).

The following R code generates a dataset with two predictors and an outcome variable that has two classes. As shown in Figure 65 (left), the two classes are separable by a linear boundary.

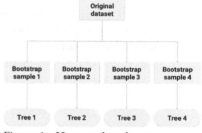

Figure 64: How random forest uses Bootstrap to grow decision trees

```
# This is a script for simulation study
rm(list = ls(all = TRUE))
require(rpart)
require(dplyr)
require(ggplot2)
```

```
require(randomForest)
ndata <- 2000
X1 <- runif(ndata, min = 0, max = 1)
X2 <- runif(ndata, min = 0, max = 1)
data <- data.frame(X1, X2)
data <- data %>% mutate(X12 = 0.5 * (X1 - X2),
                        Y = ifelse(X12 >= 0, 1, 0))
data <- data %>% select(-X12) %>% mutate(Y =
                          as.factor(as.character(Y)))
ggplot(data, aes(x = X1, y = X2, color = Y)) + geom_point() +
                labs(title = "Data points")
```

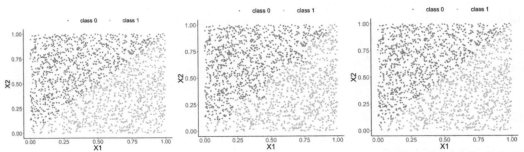

Figure 65: (Left) A linearly separable dataset with two predictors; (middle) the decision boundary of a decision tree model; (right) the decision boundary of a random forest model

Both random forest and decision tree models are applied to the dataset. The classification boundaries of both decision tree and random forest models are shown in Figures 65 (middle) and (right), respectively.

```
rf_model <- randomForest(Y ~ ., data = data)
tree_model <- rpart(Y ~ ., data = data)

pred_rf <- predict(rf_model, data, type = "prob")[, 1]
pred_tree <- predict(tree_model, data, type = "prob")[, 1]
data_pred <- data %>% mutate(pred_rf_class = ifelse(pred_rf <
  0.5, 0, 1)) %>% mutate(pred_rf_class =
  as.factor(as.character(pred_rf_class))) %>%
  mutate(pred_tree_class = ifelse(pred_tree <
  0.5, 0, 1)) %>% mutate(pred_tree_class =
                    as.factor(as.character(pred_tree_class)))
ggplot(data_pred, aes(x = X1, y = X2,
                color = pred_tree_class)) +
  geom_point() + labs(title = "Classification boundary from
                a single decision tree")
ggplot(data_pred, aes(x = X1, y = X2,
                color = pred_rf_class)) +
  geom_point() + labs(title = "Classification bounday from
                random forests")
```

We can see from Figure 65 (middle) that the classification boundary generated by the decision tree model has a difficult to approximate linear boundary. There is an inherent limitation of a tree model to fit smooth boundaries due to its box-shaped nature resulting from its use of rules to segment the data space for making predictions. In contrast, the classification boundary of the random forest model is smoother than the one of the decision tree, and it can provide better approximation of complex and nonlinear classification boundaries.

Having said that, this is not the only reason *why* the random forest model is remarkable. After all, many models can model linear boundary, and it is actually not the random forests' strength. The remarkable thing about a random forest is its capacity, as a tree-based model, to actually model linear boundary. It shows its flexibility, adaptability, and learning capacity to characterize complex patterns in a dataset. Let's see more details to understand how it works.

Theory and method

Like a decision tree, the learning process of random forests follows the algorithmic modeling framework. It uses an organized set of heuristics, rather than a mathematical characterization. We present the process of building random forest models using a simple example with a small dataset shown in Table 10 that has two predictors, x_1 and x_2, and an outcome variable with two classes.

ID	x_1	x_2	Class
1	1	1	C0
2	1	0	C1
3	0	1	C1
4	0	0	C0

Table 10: Example of a dataset

As shown in Figure 64, each tree is built on a resampled dataset that consists of data instances randomly selected from the original data set[96]. As shown in Figure 66, the first resampled dataset includes data instances (represented by their IDs) $\{1, 1, 3, 4\}$ and is used for building the first tree. The second resampled dataset includes data instances $\{2, 3, 4, 4\}$ and is used for building the second tree. This process repeats until a specific number of trees is built.

The first tree begins with the root node that contains data instances $\{1, 1, 3, 4\}$. As introduced in **Chapter 2**, we recursively split a node into two child nodes to reduce impurity (i.e., measured by entropy). This greedy recursive splitting process is also used to build each decision tree in a random forest model. A slight variation is that, in the R package `randomForest`, the **Gini index** is used to measure impurity instead of entropy.

The Gini index for a data set is defined as[97]

[96] I.e., often with the same sample size as the original dataset and is called **sampling with replacement**.

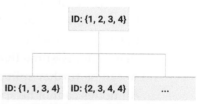

Figure 66: Examples of bootstrapped datasets from the dataset shown in Table 10

[97] C is the number of classes of the outcome variable, and p_c is the proportion of data instances that come from the class c.

$$\text{Gini} = \sum_{c=1}^{C} p_c \left(1 - p_c\right).$$

The Gini index plays the same role as the entropy (more details could be found in the Remarks section). Similar as the information gain (IG), the **Gini gain** can be defined as

$$\nabla \text{Gini} = \text{Gini}_s - \sum_{i=1,\cdots,n} w_i \text{Gini}_i.$$

Here, Gini_s is the Gini index at the node to be split; w_i and Gini_i, are the proportion of samples and the Gini index at the ith children node, respectively.

Let's go back to the first tree that begins with the root node containing data instances $\{1,1,3,4\}$. There are three instances that are associated with the class $C0$ (thus, $p_0 = \frac{3}{4}$), one instance with $C1$ (thus, $p_1 = \frac{1}{4}$). The Gini index of the root node is calculated as

$$\frac{3}{4} \times \frac{1}{4} + \frac{1}{4} \times \frac{3}{4} = 0.375.$$

To split the root node, candidates of splitting rules are:

Rule 1: $x_1 = 0$ versus $x_1 \neq 0$.

Rule 2: $x_2 = 0$ versus $x_2 \neq 0$.

The decision tree model introduced in **Chapter 2** would evaluate each of the possible splitting rules, and select the one that yields the maximum Gini gain to split the node. However, for random forests, it randomly selects the variables for splitting a node[98]. In our example, as there are two variables, we assume that x_1 is randomly selected for splitting the root node. Thus, $x_1 = 0$ is used for splitting the root node which generates the decision tree model as shown in Figure 67.

The Gini gain for the split shown in Figure 67 can be calculated as

$$0.375 - 0.5 \times 0 - 0.5 \times 0.5 = 0.125.$$

The right node in the tree shown in Figure 67 has reached a perfect state of homogeneity[99]. The left node, however, contains two instances $\{3,4\}$ that are associated with two classes. We further split the left node. Assume that this time x_2 is randomly selected. The left node can be further split as shown in Figure 68.

All nodes cannot be split further. The final tree model is shown in Figure 69, while each leaf node is labeled with the majority class of the instances in the node, such that they become decision nodes.

Applying the decision tree in Figure 69 to the 4 data points as shown in Table 10, we can get the predictions as shown in Table 11. The error rate is 25%.

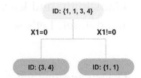

Figure 67: root node split using $x_1 = 0$

[98] In general, for a dataset with p variables, $\sqrt{p}$ variables are randomly selected for splitting.

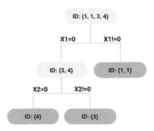

Figure 68: second split using $x_2 = 0$

[99] Which is, in practice, a rare phenomenon.

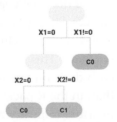

Figure 69: tree model trained

ID	x_1	x_2	Class	Prediction
1	1	1	C0	C0
2	1	0	C1	C0
3	0	1	C1	C1
4	0	0	C0	C1

Table 11: Example of a dataset

Similarly, the second, third, ..., and the mth trees can be built. Usually, in random forest models, tree pruning is not needed. Rather, we use a parameter to control the depth of the tree models to be created (i.e., use the parameter `nodesize` in `randomForest`).

When a random forest model is built, to make a prediction for a data point, each tree makes a prediction, then all the predictions are combined; e.g., for continuous outcome variable, the average of the predictions is used as the final prediction; for classification outcome variable, the class that wins majority among all trees is the final prediction.

R Lab

The 5-Step R Pipeline   **Step 1** and **Step 2** get data into your R work environment and make appropriate preprocessing.

```
# Step 1 -> Read data into R workstation
# RCurl is the R package to read csv file using a link
library(RCurl)
url <- paste0("https://raw.githubusercontent.com",
              "/analyticsbook/book/main/data/AD.csv")
AD <- read.csv(text=getURL(url))
# str(AD)
```

```
# Step 2 -> Data preprocessing
# Create your X matrix (predictors) and Y vector
# (outcome variable)
X <- AD[,2:16]
Y <- AD$DX_bl

Y <- paste0("c", Y)
Y <- as.factor(Y)

# Then, we integrate everything into a data frame
data <- data.frame(X,Y)
names(data)[16] = c("DX_bl")

# Create a training data (half the original data size)
train.ix <- sample(nrow(data),floor( nrow(data)/2) )
data.train <- data[train.ix,]
# Create a testing data (half the original data size)
data.test <- data[-train.ix,]
```

Step 3 uses the R package `randomForest` to build your random forest model.

```
# Step 3 -> Use randomForest() function to build a
# RF model
# with all predictors
library(randomForest)
rf.AD <- randomForest( DX_bl ~ ., data = data.train,
                  ntree = 100, nodesize = 20, mtry = 5)
# Three main arguments to control the complexity
# of a random forest model
```

Details for the three arguments in `randomForest` : The `ntree` is the number of trees[100]. The `nodesize` is the minimum sample size of leaf nodes[101]. The `mtry` is a parameter to control the degree of randomness when your RF model selects variables to split nodes[102].

Step 4 is prediction. We use the `predict()` function

```
# Step 4 -> Predict using your RF model
y_hat <- predict(rf.AD, data.test,type="class")
```

Step 5 evaluates the prediction performance of your model on the testing data.

```
# Step 5 -> Evaluate the prediction performance of your RF model
# Three main metrics for classification: Accuracy,
# Sensitivity (1- False Positive), Specificity (1 - False Negative)
library(caret)
confusionMatrix(y_hat, data.test$DX_bl)
```

The result is shown below. It is an information-rich object[103].

```
## Confusion Matrix and Statistics
##
##             Reference
## Prediction  c0  c1
##         c0 136  31
##         c1   4  88
##
##              Accuracy : 0.8649
##                95% CI : (0.8171, 0.904)
##   No Information Rate : 0.5405
##   P-Value [Acc > NIR] : < 2.2e-16
##
##                 Kappa : 0.7232
##
## Mcnemar's Test P-Value : 1.109e-05
##
##           Sensitivity : 0.9714
```

[100] The more trees, the more complex the random forest model is.
[101] The larger the sample size in leaf nodes, the less depth of the trees; therefore, the less complex the random forest model is.
[102] For classification, the default value of `mtry` is $\sqrt{p}$, where p is the number of variables; for regression, the default value of `mtry` is $p/3$.

[103] To learn more about an R object, function, and package, please check out the online documentation that is usually available for an R package that has been released to the public. For example, for the `confusionMatrix` in the R package `caret` , check out this link: https://www.rdocumentation.org/packages/caret/versions/6.0-84/topics/confusionMatrix. Some R packages also come with a journal article published in the *Journal of Statistical Software*. E.g., for `caret` , see Kuhn, M., *Building predictive models in R using the caret package*, Journal of Statistical Software, Volume 28, Issue 5, 2018, http://www.jstatsoft.org/article/view/v028i05/v28i05.pdf.

```
##            Specificity : 0.7395
##         Pos Pred Value : 0.8144
##         Neg Pred Value : 0.9565
##            Prevalence : 0.5405
##        Detection Rate : 0.5251
## Detection Prevalence : 0.6448
##     Balanced Accuracy : 0.8555
##
##        'Positive' Class : c0
```

We can also draw the ROC curve

```
# ROC curve is another commonly reported metric
# for classification models
library(pROC)
# pROC has the roc() function that is very useful here
y_hat <- predict(rf.AD, data.test,type="vote")
# In order to draw ROC, we need the intermediate prediction
# (before RF model binarize it into binary classification).
# Thus, by specifying the argument type="vote", we can
# generate this intermediate prediction. y_hat now has
# two columns, one corresponds to the ratio of votes the
# trees assign to one class, and the other column is the
# ratio of votes the trees assign to another class.
main = "ROC Curve"
plot(roc(data.test$DX_bl, y_hat[,1]),
     col="blue", main=main)
```

And we can have the ROC curve as shown in Figure 70.

Beyond the 5-Step R Pipeline   Random forests are complex models that have many parameters to be tuned. In **Chapter 2** and **Chapter 3** we have used the `step()` function for automatic model selection for regression models. Part of the reason this is possible for regression models is that model selection for regression models largely concerns variable selection only. For decision tree and random forest models, the model selection concerns not only variable selection, but also many other aspects, such as the depth of the tree, the number of trees, and the degree of randomness that would be used in model training. This makes the model selection for tree models a craft. An individual's experience and insights make a difference, and some may find a better model, even the same package is used on the same dataset to build models[104].

To see how these parameters impact the models, we conduct some experiments. The number of trees is one of the most important parameters of random forests that we'd like to be tuned well. We can build different random forest models by tuning the parameter `ntree` in `randomForest`. For each selection of the number

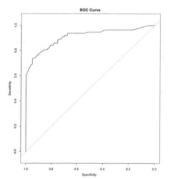

Figure 70: ROC curve of the RF model

[104] There is often an impression that a good model is built by a good pipeline, like this 5-step pipeline. This impression is a reductive view, since it only looks at the final stage of data analytics. Like manufacturing, when the process is mature and we are able to see the rationale behind every detail of the manufacturing process, we may lose sight of those alternatives that had been considered, experimented, then discarded (or withheld) for various reasons.

of trees, we first randomly split the dataset into training and testing datasets, then train the model on the training dataset, and evaluate its performance on the testing dataset. This process of data splitting, model training, and testing is repeated 100 times. We can use boxplots to show the overall performance of the models. Results are shown in Figure 71.

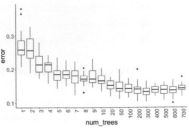

Figure 71: Error v.s. number of trees in a random forest model

```r
library(rpart)
library(dplyr)
library(tidyr)
library(ggplot2)
require(randomForest)
library(RCurl)
set.seed(1)

theme_set(theme_gray(base_size = 15))
url <- paste0("https://raw.githubusercontent.com",
             "/analyticsbook/book/main/data/AD.csv")
data <- read.csv(text=getURL(url))

target_indx <- which(colnames(data) == "DX_bl")
data[, target_indx] <-
  as.factor(paste0("c", data[, target_indx]))
rm_indx <- which(colnames(data) %in% c("ID", "TOTAL13",
                                        "MMSCORE"))
data <- data[, -rm_indx]
results <- NULL
for (itree in c(1:9, 10, 20, 50, 100, 200, 300, 400, 500,
    600, 700)) {
for (i in 1:100) {
train.ix <- sample(nrow(data), floor(nrow(data)/2))
rf <- randomForest(DX_bl ~ ., ntree = itree, data =
                                    data[train.ix, ])
pred.test <- predict(rf, data[-train.ix, ], type = "class")
this.err <- length(which(pred.test !=
            data[-train.ix, ]$DX_bl))/length(pred.test)
results <- rbind(results, c(itree, this.err))
}
}

colnames(results) <- c("num_trees", "error")
results <- as.data.frame(results) %>%
  mutate(num_trees = as.character(num_trees))
levels(results$num_trees) <- unique(results$num_trees)
results$num_trees <- factor(results$num_trees,
                        unique(results$num_trees))
ggplot() + geom_boxplot(data = results, aes(y = error,
            x = num_trees)) + geom_point(size = 3)
```

It can be seen in Figure 71 that when the number of trees is small, particularly, less than 10, the improvement on prediction performance of random forest is substantial with trees added. However, the error rates become stable after the number of trees reaches 100.

Next, let's consider the number of features (i.e., use the parameter `mtry` in the function `randomForest`). Here, 100 trees are used. For each number of features, again, following the process we have used in the experiment with the number of trees, we draw the boxplots in Figure 72.

It can be seen that the error rates are not significantly different when the number of features changes.

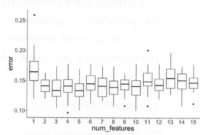

Figure 72: Error v.s. number of features in a random forest model

```r
library(rpart)
library(dplyr)
library(tidyr)
library(ggplot2)
require(randomForest)
library(RCurl)
set.seed(1)
theme_set(theme_gray(base_size = 15))

url <- paste0("https://raw.githubusercontent.com",
              "/analyticsbook/book/main/data/AD.csv")
data <- read.csv(text=getURL(url))

target_indx <- which(colnames(data) == "DX_bl")
data[, target_indx] <- as.factor(
        paste0("c", data[, target_indx]))
rm_indx <- which(colnames(data) %in% c("ID", "TOTAL13",
                                       "MMSCORE"))
data <- data[, -rm_indx]
nFea <- ncol(data) - 1
results <- NULL
for (iFeatures in 1:nFea) {
for (i in 1:100) {
train.ix <- sample(nrow(data), floor(nrow(data)/2))
rf <- randomForest(DX_bl ~ ., mtry = iFeatures, ntree = 100,
                   data = data[train.ix,])
pred.test <- predict(rf, data[-train.ix, ], type = "class")
this.err <- length(which(pred.test !=
                data[-train.ix, ]$DX_bl))/length(pred.test)
results <- rbind(results, c(iFeatures, this.err))
}
}
colnames(results) <- c("num_features", "error")
results <- as.data.frame(results) %>%
  mutate(num_features = as.character(num_features))
levels(results$num_features) <- unique(results$num_features)
results$num_features <- factor(results$num_features,
                               unique(results$num_features))
ggplot() + geom_boxplot(data = results, aes(y = error,
                x = num_features)) + geom_point(size = 3)
```

Further, we experiment with the minimum node size (i.e., use the parameter `nodesize` in the function `randomForest`), that is, the minimum number of instances at a node. Boxplots of their performances are shown in Figure 73.

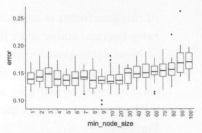

Figure 73: Error v.s. node size in a random forest model

```r
library(dplyr)
library(tidyr)
library(ggplot2)
require(randomForest)
library(RCurl)
set.seed(1)

theme_set(theme_gray(base_size = 15))

url <- paste0("https://raw.githubusercontent.com",
              "/analyticsbook/book/main/data/AD.csv")

data <- read.csv(text=getURL(url))

target_indx <- which(colnames(data) == "DX_bl")
data[, target_indx] <- as.factor(paste0("c", data[, target_indx]))
rm_indx <- which(colnames(data) %in% c("ID", "TOTAL13",
                                       "MMSCORE"))
data <- data[, -rm_indx]

results <- NULL
for (inodesize in c(1, 2, 3, 4, 5, 6, 7, 8, 9, 10, 20, 30,
    40, 50, 60, 70, 80,90, 100)) {
  for (i in 1:100) {
    train.ix <- sample(nrow(data), floor(nrow(data)/2))
    rf <- randomForest(DX_bl ~ ., ntree = 100, nodesize =
                       inodesize, data = data[train.ix,])
    pred.test <- predict(rf, data[-train.ix, ], type = "class")
    this.err <- length(which(pred.test !=
                      data[-train.ix, ]$DX_bl))/length(pred.test)
    results <- rbind(results, c(inodesize, this.err))
    # err.rf <- c(err.rf, length(which(pred.test !=
    # data[-train.ix,]$DX_bl))/length(pred.test) )
  }
}

colnames(results) <- c("min_node_size", "error")
results <- as.data.frame(results) %>%
  mutate(min_node_size = as.character(min_node_size))
levels(results$min_node_size) <- unique(results$min_node_size)
results$min_node_size <- factor(results$min_node_size,
                                unique(results$min_node_size))
ggplot() + geom_boxplot(data = results, aes(y = error,
                x = min_node_size)) + geom_point(size = 3)
```

Figure 73 shows that the error rates start to rise when the minimum node size equals 40. And the error rates are not substantially different when the minimum node size is less than 40. All together, these results provide information for us to select models.

To compare random forest with decision tree, we can also follow a similar process, i.e., half of the dataset is used for training and the other half for testing. This process of splitting data, training the model on training data, and testing the model on testing data is repeated 100 times, and boxplots of the errors from decision trees and random forests are plotted in Figure 74 using the following R code.

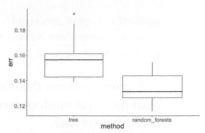

Figure 74: Performance of random forest v.s. tree model on the Alzheimer's disease data

```r
library(rpart)
library(dplyr)
library(tidyr)
library(ggplot2)
library(RCurl)
require(randomForest)
set.seed(1)

theme_set(theme_gray(base_size = 15))

url <- paste0("https://raw.githubusercontent.com",
              "/analyticsbook/book/main/data/AD.csv")
data <- read.csv(text=getURL(url))

target_indx <- which(colnames(data) == "DX_bl")
data[, target_indx] <- as.factor(paste0("c", data[, target_indx]))
rm_indx <- which(colnames(data) %in%
                  c("ID", "TOTAL13", "MMSCORE"))
data <- data[, -rm_indx]

err.tree <- NULL
err.rf <- NULL
for (i in 1:100) {
train.ix <- sample(nrow(data), floor(nrow(data)/2))
tree <- rpart(DX_bl ~ ., data = data[train.ix, ])
pred.test <- predict(tree, data[-train.ix, ], type = "class")
err.tree <- c(err.tree, length(
  which(pred.test != data[-train.ix, ]$DX_bl))/length(pred.test))

rf <- randomForest(DX_bl ~ ., data = data[train.ix, ])
pred.test <- predict(rf, data[-train.ix, ], type = "class")
err.rf <- c(err.rf, length(
  which(pred.test != data[-train.ix, ]$DX_bl))/length(pred.test))
}
err.tree <- data.frame(err = err.tree, method = "tree")
err.rf <- data.frame(err = err.rf, method = "random_forests")
```

```
ggplot() + geom_boxplot(
  data = rbind(err.tree, err.rf), aes(y = err, x = method)) +
geom_point(size = 3)
```

Figure 74 shows that the error rates of decision trees are higher than those for random forests, indicating that random forest is a better model here.

Remarks

The Gini index versus the entropy

The *Gini index* plays the same role as the *entropy*. To see that, the following R code plots the Gini index and the entropy in a binary class problem[105]. Their similarity is evident as shown in Figure 75.

The following R scripts package the Gini index and entropy as two functions.

[105] A binary class problem has two nominal parameters, p_1 and p_2. Because p_2 equals $1 - p_1$, it is actually a one-parameter system, such that we can visualize how the Gini index and the entropy changes according to p_1, as shown in Figure 75.

```
entropy <- function(p_v) {
e <- 0
for (p in p_v) {
if (p == 0) {
this_term <- 0
} else {
this_term <- -p * log2(p)
}
e <- e + this_term
}
return(e)
}
```

```
gini <- function(p_v) {
e <- 0
for (p in p_v) {
if (p == 0) {
this.term <- 0
} else {
this.term <- p * (1 - p)
}
e <- e + this.term
}
return(e)
}
```

The following R script draws Figure 75.

```
entropy.v <- NULL
gini.v <- NULL
p.v <- seq(0, 1, by = 0.01)
```

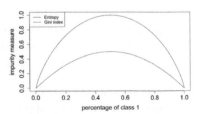

Figure 75: Gini index vs. entropy

```
for (p in p.v) {
entropy.v <- c(entropy.v, (entropy(c(p, 1 - p))))
gini.v <- c(gini.v, (gini(c(p, 1 - p))))
}
plot(p.v, gini.v, type = "l", ylim = c(0, 1),
    xlab = "percentage of class 1",col = "red",
    ylab = "impurity measure", cex.lab = 1.5,
    cex.axis = 1.5, cex.main = 1.5,cex.sub = 1.5)
lines(p.v, entropy.v, col = "blue")
legend("topleft", legend = c("Entropy", "Gini index"),
       col = c("blue", "red"), lty = c(1, 1), cex = 0.8)
```

It can be seen in Figure 75 that the two impurity measures are similar. Both reach minimum, i.e., 0, when all the data instances belong to the same class, and they reach maximum when there are equal numbers of data instances for the two classes. In practice, they produce similar trees.

Why random forests work

A random forest model is inspirational because it shows that *randomness*, usually considered as a troublemaker, has a productive dimension. This seems to be counterintuitive. An explanation has been pointed out in numerous literature that the random forests, together with other models that are called **ensemble learning** models, could make a group of **weak models** come together to form a strong model.

For example, consider a random forest model with 100 trees. Each tree is a *weak model* and its accuracy is 0.6. Assume that the trees are independent[106], with 100 trees the probability of the random forest model to predict correctly on any data point is 0.97, i.e.,

$$\sum_{k=51}^{100} C(n,k) \times 0.6^k \times 0.4^{100-k} = 0.97.$$

This result is impressive, but don't forget the *assumption of the independence* between the trees. This assumption does not hold in reality in a strict sense; i.e., ideally, we hope to have an algorithm that can find many good models that perform well and are all different; but, if we build many models using one dataset, these models would more or less resemble each other. Particularly, when we solely focus on models that can achieve optimal performance, it is often that the identified models end up more or less the same. Responding to this dilemma, randomness (i.e., the use of Bootstrap to randomize choices of data instances and the use of random feature selection for building trees) is introduced into the model-building process to create diversity of the models. The dynamics between the degree of randomness, the performance of each individual model, and their

[106] I.e., the predictions of one tree provide no hint to guess the predictions of another tree.

difference, should be handled well. To develop the craft, you may have many practices and focus on driving this dynamics towards a collective good.

Variable importance by random forests

Recall that a random forest model consists of decision trees that are defined by splits on some variables. The splits provide information about variables' importance. To explain this, consider the data example shown in Table 12.

ID	x_1	x_2	x_3	x_4	Class
1	1	1	0	1	C0
2	0	0	0	1	C1
3	1	1	1	1	C1
4	0	0	1	1	C1

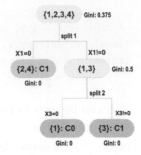

Figure 76: Tree #1

Table 12: Example of a dataset

Assume that a random forest model with two trees (i.e., shown in Figures 76 and 77) is built on the dataset.

At split 1, the Gini gain for x_1 is:

$$0.375 - 0.5 \times 0 - 0.5 \times 0.5 = 0.125.$$

At split 2, the Gini gain for x_3 is:

$$0.5 - 0.5 \times 0 - 0.5 \times 0 = 0.5.$$

At split 3, the Gini gain for x_2 is

$$0.5 - 0.25 \times 0 + 0.75 \times 0.44 = 0.17.$$

At split 4, the Gini gain for x_3 is

$$0.44 - 0.5 \times 0 + 0.5 \times 0 = 0.44.$$

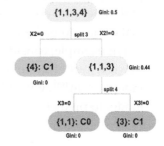

Figure 77: Tree #2

Table 13 summarizes the contributions of the variables in the splits. The total contribution can be used as the variable's importance score.

ID of splits	x_1	x_2	x_3	x_4
1	0.125	0	0	0
2	0	0	0.5	0
3	0	0.17	0	0
4	0	0	0.44	0
Total	0.125	0.17	0.94	0

Table 13: Contributions of the variables in the splits

The variable's importance score helps us identify the variables that have strong predictive values. The approach to obtain the variable's importance score is simple and effective, but it is not perfect. For example, if we revisit the example shown in Table 12, we may notice that x_1 and x_2 are identical. What does this suggest to you[107]?

Partial dependency plot

Variable importance scores indicate whether a variable is informative in predicting the outcome variable. It does not provide information about *how* the outcome variable is influenced by the variables. **Partial dependency plot** can be used to visualize the relationship between a predictor and the outcome variable, averaged on other predictors.

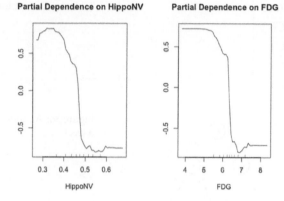

[107] Interested readers may read this article: Deng, H. and Runger, G., *Gene selection with guided regularized random forest*, Pattern Recognition, Volume 46, Issue 12, Pages 3483-3489, 2013.

Figure 78: Partial dependency plots of variables in random forests

We draw in Figure 78 the partial dependency plots of two variables on the AD dataset. It is clear that the relationships between the outcome variable with both predictors are significant. And the orientation of both relationships is visualized by the plots.

```
# Draw the partial dependency plots of variables in random forests
randomForest::partialPlot(rf, data, HippoNV, "1")
randomForest::partialPlot(rf, data, FDG, "1")
```

Exercises

1. Continue the example in the 4-step R pipeline R lab in this chapter that estimated the mean and standard derivation of the variable `HippoNV` of the AD dataset. Use the same R pipeline to evaluate the uncertainty of the estimation of the standard derivation of the variable `HippoNV` of the AD dataset. Report its 95% CI.

2. Use the R pipeline for Bootstrap to evaluate the uncertainty of the estimation of the coefficient of a logistic regression model. Report

the 95% CI of the estimated coefficients.

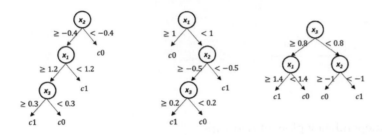

Figure 79: A random forest model with 2 trees

3. Consider the data in Table 14. Assume that two trees were built on it, as shown in Figure 79. (a) Calculate the Gini index of each node of both trees; and (b) estimate the importance scores of the three variables in this RF model.

ID	x_1	x_2	x_3	Class
1	1	0	1	C0
2	0	0	1	C1
3	1	1	1	C1
4	0	1	1	C1

Table 14: Dataset for Q3

Figure 80: A random forest model with 3 trees

4. A random forest model with 3 trees is built, as shown in Figure 80. Use this model to predict on the data in Table 15.

5. Follow up on the simulation experiment in Q9 in Chapter 2. Apply random forest with 100 trees (by setting `ntree = 100`) on the simulated data and comment on the result.

ID	x_1	x_2	x_3	Class
1	2	−0.5	0.5	
2	0.8	−1.1	0.1	
3	1.2	−0.3	0.9	

Table 15: Dataset for Q4

Chapter 5: Learning (I)
Cross-validation & OOB

Overview

The question of *learning* in data analytics concerns *whether or not the model has learned from the data*. To understand this, let's look at a few dilemmas.

Dilemma 1 Let's consider the prediction of a disease. Hired by the Centers of Disease Control and Prevention (CDC), a data scientist built a prediction model (e.g., a logistic regression model) using tens of thousands of patient' data collected over several years, and the model's prediction accuracy was 90%. Isn't this a good model?

Then we are informed that this is a rare disease, and national statistics has shown that only 0.001% of the population of the United States have this disease. This contextual knowledge changes our perception of the 90% prediction accuracy dramatically. Consider a trivial model that simply predicts all the cases as negative (i.e., no disease), wouldn't this trivial model achieve a prediction accuracy as high as 99.999%?[108]

[108] *Moral of the story:* context matters.

Dilemma 2 Now let's look at another example. Some studies pointed out that bestsellers could be reasonably predicted by computers based on the book's content. These studies collected a number of books, some were bestsellers (i.e., based on *The New York Times* book-selling rank). They extracted features from these books, such as some thematic features and linguistic patterns that could be measured by words use and frequency, and trained prediction models with these features as predictors and the bestseller/non-bestseller as a binary outcome. The model achieved an accuracy between 70% to 80%. This looks promising. The problem is that, e.g., in the dataset, you may see that about 30% of the books were bestsellers.

Statistics show that in recent years there could have been more than one million new books published each year. How many of them would make *The New York Times* bestseller list? Maybe 0.01% or fewer. So the dataset collected for training the data is not entirely representative of the population. And in fact, this is another rare-disease-like situation[109].

[109] *Moral of the story:* how the dataset is collected also matters.

Dilemma 3   As we have mentioned in **Chapter 2**, the *R-squared* measures the goodness-of-fit of a regression model. Let's revisit the definition of the R-squared

$$\text{R-squared} = \frac{\sigma_y^2 - \sigma_\varepsilon^2}{\sigma_y^2}.$$

We can see that the denominator is always fixed, no matter how we change the regression model; while the numerator could only decrease if more variables are put into the model, even if these new variables have no relationship with the outcome variable. In other words, with more variables in the model, even if the residuals are not reduced, at worst they remain the same[110].

[110] For a more formal discussions about the technical limitations of R-squared, e.g., a good starting point is: Kvalseth, T.O., *Cautionary Note about R^2*, The American Statistician, Volume 39, Issue 4, Pages 279-285, 1985.

Further, the *R-squared* is impacted by the variance of the predictors as well. As the regression model is

$$y = x\beta + \epsilon,$$

it is known that the variance of y is the variance of $\text{var}(y) = \beta^T \text{var}(x)\beta + \text{var}(\epsilon)$. The R-squared can be rewritten as

$$R^2 = \frac{\beta^T \text{var}(x)\beta}{\beta^T \text{var}(x)\beta + \text{var}(\epsilon)}.$$

Thus, the *R-squared* is not only impacted by how well x can predict y, but also by the variance of x as well[111].

[111] *Moral of the story:* a model's performance metric could be manipulated under "legal" terms.

Cross-validation

Rationale and formulation

Performance metrics, such as accuracy and R-squared, are context-dependent (*Dilemma 1*), data-dependent (*Dilemma 2*), and vulnerable to conscious or unconscious manipulations (*Dilemma 3*). These limitations make them *relative* metrics. They are not the *absolutes* that we can rely on to evaluate models in a universal fashion in all contexts.

So, what should be the universal and objective criteria to evaluate the learning performance of a model?

To answer this question, we need to understand the concepts, **underfit, good fit**, and **overfit**.

Figure 81 shows three models to fit the same dataset that has two classes of data points. The first model is a linear model[112] that

[112] E.g., $f_1(x) = \beta_0 + \beta_1 x_1 + \beta_2 x_2$.

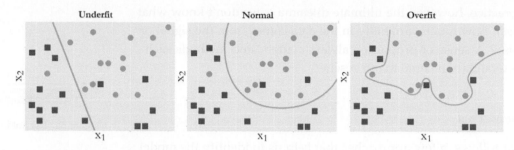

Figure 81: Three types of model performance

yields a straight line as the **decision boundary**. Obviously, many data points are misclassified when using a linear decision boundary. Some curvature is needed to bend the decision boundary, so we introduce some second order terms and an interaction term of the two predictors to create another model[113]. The decision boundary is shown in Figure 81 (middle). This improved model still could not classify the two classes completely. More interaction terms[114] are introduced into the model. The decision boundary is shown in Figure 81 (right).

Now 100% prediction accuracy could be achieved. A sense of suspicion should arise: is this *too good to be true*?

What we have seen in Figure 81, on the positive side, is the capacity we can develop to *fit* a dataset[115]. On the other hand, what is responsible for the sense of suspicion of "too good to be true" is that we didn't see a *validation process* at work.

Recall a general assumption of data modeling is[116]

$$\underbrace{y}_{data} = \underbrace{f(x)}_{signal} + \underbrace{\epsilon}_{noise}$$

where *noise* is unpredictable. Bearing this framework in mind, we revisit the three models in Figure 81, which from left to right illustrate **underfit**, **good fit**, and **overfit**, respectively. A model called *underfitted* means it fails to incorporate some pattern of the signal in the dataset. A model called *overfitted* means it allows the noise to affect the model[117]. A dataset could be randomly generated, but the *mechanism of generating the randomness*[118] is a constancy. The model in the middle panel of Figure 81 is able to maintain a balance: it captures the structural constancy in the data to form the model, while resisting the noise and refusing to let them bend its decision boundary.

In summary, Figure 81 illustrates that:

- **Overfit**: Complexity of the model > complexity of the signal;

- **Good fit**: Complexity of the model = complexity of the signal;

- **Underfit**: Complexity of the model < complexity of the signal.

[113] E.g., $f_2(x) = \beta_0 + \beta_1 x_1 + \beta_2 x_2 + \beta_{11} x_1^2 + \beta_{22} x_2^2 + \beta_{12} x_1 x_2$.

[114] E.g., $f_3(x) = \beta_0 + \beta_1 x_1 + \beta_2 x_2 + \beta_{11} x_1^2 + \beta_{22} x_2^2 + \beta_{12} x_1 x_2 + \beta_{112} x_1^2 x_2 + \beta_{122} x_1 x_2^2 + \cdots$.

[115] *Fit* a dataset is not necessarily *model* a dataset. Beginners may need time to develop a sense to see the difference between the two.

[116] I.e., Eq. 2 in **Chapter 2**.

[117] Noise, by definition, only happens by accident. While the model, by definition, is to generalize the constancy, i.e., the signal, of the data rather than its unrepeatable randomness.

[118] I.e., like a distribution model.

In practice, however, the ultimate dilemma is we don't know what to expect: how much variability in the data comes from the signal or the noise? A sense of proportion always matters, and methods such as cross-validation come to our rescue.

Theory/Method

In what follows, a few approaches that help us to identify the model with *good fit* (i.e., the one shown in Figure 81 (middle), $f_2(x)$) are introduced. These approaches share the same goal: to train a model on the training data and make sure the learned model would succeed on an *unseen* testing dataset.

The first approach is the **hold-out** method. As shown in Figure 82, the hold-out method randomly divides a given dataset into two parts. The model is trained on the training data only, while its performance is evaluated on the testing data. For instance, for the three models shown in Figure 81, each of them will be trained on the training dataset and will have their regression coefficients estimated. Then, the learned models will be evaluated on the testing data. The model that has the best performance on the testing data will be selected as the final model.

Another approach called **random sampling** repeats this random division many times, as shown in Figure 83. Each time, the *model training and selection* only uses the training dataset, and the model evaluation only uses the testing dataset. The performance of the models on the three experiments could be averaged and the model that has the best average performance is selected.

The **K-fold cross-validation** is a mix of the *random sampling* method and the *hold-out* method. It first divides the dataset into K folds of equal sizes. Then, it trains a model using any combination of $K - 1$ folds of the dataset, and tests the model using the remaining one-fold of the dataset. As shown in Figure 84, the model training and testing process is repeated K times. The performance of the models on the K experiments could be averaged and the model that has the best average performance is selected.

These approaches can be used for evaluating a model's performance in a robust way. They are also useful when we'd like to choose among model types[119] or model formulations[120]. While the model type and the model formulation is settled, for example, suppose that we have determined to use linear regression and the model formulation $y = \beta_0 + \beta_1 x_1 + \beta_2 x_2$, these methods could be used to evaluate the performance of this single model. It is not uncommon that in real data analysis, these cross-validation and sampling methods are used in combination and serve different stages of the analysis process.

Figure 82: The hold-out method

Figure 83: The random sampling method

Figure 84: The K-fold cross-validation method (here, $K = 4$)

[119] E.g., decision tree vs. linear regression.

[120] E.g., model 1: $y = \beta_0 + \beta_1 x_1$; vs. model 2: $y = \beta_0 + \beta_1 x_1 + \beta_2 x_2$.

R Lab

The 4-Step R Pipeline   **Step 1** and **Step 2** are standard procedures to get data into R and further make appropriate preprocessing.

```
# Step 1 -> Read data into R workstation

library(RCurl)
url <- paste0("https://raw.githubusercontent.com",
              "/analyticsbook/book/main/data/AD.csv")
AD <- read.csv(text=getURL(url))

# Step 2 -> Data preprocessing
# Create your X matrix (predictors) and Y vector
# (outcome variable)
X <- AD[,2:16]
Y <- AD$MMSCORE
data <- data.frame(X,Y)
names(data)[16] <- c("MMSCORE")
```

Step 3 creates a list of models to be evaluated and compared with[121].

```
# Step 3 -> gather a list of candidate models
# Use linear regression model as an example
model1 <- "MMSCORE ~ ."
model2 <- "MMSCORE ~ AGE + PTEDUCAT + FDG + AV45 + HippoNV +
                                        rs3865444"
model3 <- "MMSCORE ~ AGE + PTEDUCAT"
model4 <- "MMSCORE ~ FDG + AV45 + HippoNV"
```

[121] *Linear regression*: we often compare models using different predictors;
Decision tree: we often compare models with different depths;
Random forests: we often compare models with a different number of trees, a different depth of individual trees, or a different number of features to be randomly picked up to split the nodes.

Step 4 uses the 10-fold cross-validation to evaluate the models and find out which one is the best. The R code is shown below and is divided into two parts. The first part uses the `sample()` function to create random split of the dataset into 10 folds.

```
# Step 4 -> Use 10-fold cross-validation to evaluate all models

# First, let me use 10-fold cross-validation to evaluate the
# performance of model1
n_folds = 10
# number of fold (the parameter K in K-fold cross validation)
N <- dim(data)[1] # the sample size, N, of the dataset
folds_i <- sample(rep(1:n_folds, length.out = N))
# This randomly creates a labeling vector (1 X N) for
# the N samples. For example, here, N = 16, and
# I run this function and it returns
# the value as 5 4 4 10 6 7 6 8 3 2 1 5 3 9 2 1.
# That means, the first sample is allocated to the 5th fold,
# the 2nd and 3rd samples are allocated to the 4th fold, etc.
```

The second part shows how we evaluate the models. We only show the code for two models, as the script for evaluating each model is basically the same.

```r
# Evaluate model1
# cv_mse aims to make records of the mean squared error
# (MSE) for each fold
cv_mse <- NULL
for (k in 1:n_folds) {
  test_i <- which(folds_i == k)
  # In each iteration of the 10 iterations, remember, we use one
  # fold of data as the testing data
  data.train <- data[-test_i, ]
  # Then, the remaining 9 folds' data form our training data
  data.test <- data[test_i, ]
  # This is the testing data, from the ith fold
  lm.AD <- lm(model1, data = data.train)
  # Fit the linear model with the training data
  y_hat <- predict(lm.AD, data.test)
  # Predict on the testing data using the trained model
  true_y <- data.test$MMSCORE
  # get the true y values for the testing data
  cv_mse[k] <- mean((true_y - y_hat)^2)
  # mean((true_y - y_hat)^2): mean squared error (MSE).
  # The smaller this error, the better your model is
}
mean(cv_mse)


# Then, evaluate model2
cv_mse <- NULL
# cv_mse aims to make records of the mean squared error (MSE)
# for each fold
for (k in 1:n_folds) {
  test_i <- which(folds_i == k)
  # In each iteration of the 10 iterations, remember,
  # we use one fold of data as the testing data
  data.train <- data[-test_i, ]
  # Then, the remaining 9 folds' data form our training data
  data.test <- data[test_i, ]
  # This is the testing data, from the ith fold
  lm.AD <- lm(model2, data = data.train)
  # Fit the linear model with the training data
  y_hat <- predict(lm.AD, data.test)
  # Predict on the testing data using the trained model
  true_y <- data.test$MMSCORE
  # get the true y values for the testing data
  cv_mse[k] <- mean((true_y - y_hat)^2)
  # mean((true_y - y_hat)^2): mean squared error (MSE).
  # The smaller this error, the better your model is
}
```

```
mean(cv_mse)

# Then, evaluate model3 ...
# Then, evaluate model4 ...
```

The result is shown below.

```
## [1] 3.17607
## [1] 3.12529
## [1] 4.287637
## [1] 3.337222
```

We conclude that `model2` is the best one, as it achieves the minimum mean squared error (MSE).

Simulation Experiment   How do we know the cross-validation could identify a good model, i.e., the one that neither overfits nor underfits the data? Let's design a simulation experiment to study the performance of cross-validation[122].

The purpose of the experiment is two-fold: (1) to show that the cross-validation can help us mitigate the model selection problem, and (2) to show that R is not just a tool for implementing data analysis methods, but also an experimental tool to gain first-hand experience of any method's practical performance.

Our experiment has a clearly defined metric to measure the complexity of the *signal*. We resort to the **spline** models[123] that could be loosely put into the category of regression models, which have a precise mechanism to tune a model's complexity, i.e., through the parameter of **degree of freedom** (**df**). For simplicity, we simulate a dataset with one predictor and one outcome variable. In R, we use the `ns()` function to simulate the spline model.

The outcome is a nonlinear curve[124]. We use the degree of freedom (`df`) parameter in the `ns()` function to control the complexity of the curve, i.e., the larger the `df`, the more "nonlinear" the curve. As this curve is the *signal* of the data, we also simulate noise through a Gaussian distribution using the `rnorm()` function.

```
# Write a simulator to generate dataset with one predictor and
# one outcome from a polynomial regression model
seed <- rnorm(1)
set.seed(seed)
gen_data <- function(n, coef, v_noise) {
eps <- rnorm(n, 0, v_noise)
x <- sort(runif(n, 0, 100))
X <- cbind(1,ns(x, df = (length(coef) - 1)))
y <- as.numeric(X %*% coef + eps)
return(data.frame(x = x, y = y))    }
```

[122] A large portion of the R script in this subsection was modified from `malanor.net`, now no longer an active site.

[123] A good tutorial: Eilers, P. and Marx, B., *Splines, Knots, and Penalties*, Computational statistics, Volume 2, Issue 6, Pages 637-653, 2010.

[124] Here, we use the B-spline basis matrix for natural cubic splines to create a nonlinear curve. This topic is beyond the scope of this book.

The following R codes generate the scattered grey data points and the true model as shown in Figure 85.

```r
# install.packages("splines")
require(splines)
## Loading required package: splines
# Simulate one batch of data, and see how different model
# fits with df from 1 to 50

n_train <- 100
coef <- c(-0.68,0.82,-0.417,0.32,-0.68)
v_noise <- 0.2
n_df <- 20
df <- 1:n_df
tempData <- gen_data(n_train, coef, v_noise)

x <- tempData[, "x"]
y <- tempData[, "y"]
# Plot the data
x <- tempData$x
X <- cbind(1, ns(x, df = (length(coef) - 1)))
y <- tempData$y
plot(y ~ x, col = "gray", lwd = 2)
lines(x, X %*% coef, lwd = 3, col = "black")
```

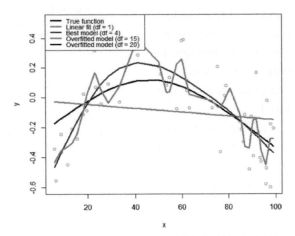

Figure 85: The simulated data from a nonlinear regression model with B-spline basis matrix (df =4), and various fitted models with different degrees of freedom

We then fit the data with a variety of models, starting from df=1 [125] to df=20 [126]. The fitted curves are overlaid onto the scattered data points in Figure 85. It can be seen that the linear model obviously underfits the data, as it lacks the flexibility to characterize the complexity of the signal sufficiently. The model that has (df=20) overfits the data, evidenced by its complex shape. It tries too hard to fit the local patterns, i.e., by all the turns and twists of its curve,

[125] I.e., corresponds to the linear model.
[126] I.e., a very complex model.

while the local patterns were mostly induced by noise[127].

```
# Fit the data using different models with different
# degrees of freedom (df)
fit <- apply(t(df), 2, function(degf) lm(y ~ ns(x, df = degf)))
# Plot the models
plot(y ~ x, col = "gray", lwd = 2)
lines(x, fitted(fit[[1]]), lwd = 3, col = "darkorange")
lines(x, fitted(fit[[4]]), lwd = 3, col = "dodgerblue4")
# lines(x, fitted(fit[[10]]), lwd = 3, col = "darkorange")
lines(x, fitted(fit[[20]]), lwd = 3, col = "forestgreen")
legend(x = "topright", legend = c("True function",
     "Linear fit (df = 1)", "Best model (df = 4)",
     "Overfitted model (df = 15)","Overfitted model (df = 20)"),
     lwd = rep(3, 4), col = c("black", "darkorange", "dodgerblue4",
     "forestgreen"), text.width = 32, cex = 0.6)
```

Note that, in this example, we have known that the true model has `df=4`. In reality, we don't have this knowledge. It is dangerous to keep increasing the model complexity to aggressively pursue better prediction performance *on the training data*. To see the danger, let's do another experiment.

First, we use the following R code to generate a testing data from the same distribution of the training data.

```
# Generate test data from the same model
n_test <- 50
xy_test <- gen_data(n_test, coef, v_noise)
```

Then, we fit a set of models from linear (`df=1`) to (`df=20`) using the training dataset. And we compute the prediction errors of these models using the training dataset and testing dataset separately. This is done by the following R script.

```
# Compute the training and testing errors for each model
mse <- sapply(fit, function(obj) deviance(obj)/nobs(obj))
pred <- mapply(function(obj, degf) predict(obj, data.frame(x =
                        xy_test$x)),fit, df)
te <- sapply(as.list(data.frame(pred)),
          function(y_hat) mean((xy_test$y - y_hat)^2))
```

We further present the training and testing errors of the models in Figure 86, by running the R script below.

[127] A model that tries too hard to fit the training data by absorbing its noise into its shape will not perform well on future unseen testing data, since the particular noise in the training data would not appear in the testing data—if a noise repeats itself, it is not noise anymore but signal.

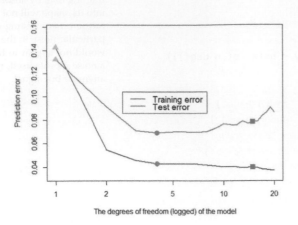

Figure 86: Prediction errors of the models (from (`df` = 0) to (`df` = 20)) on the training dataset and testing data

```
# Plot the errors
plot(df, mse, type = "l", lwd = 2, col = gray(0.4),
    ylab = "Prediction error",
    xlab = "The degrees of freedom (logged) of the model",
    ylim = c(0.9*min(mse), 1.1*max(mse)), log = "x")

lines(df, te, lwd = 2, col = "orange3")

points(df[1], mse[1], col = "palegreen3", pch = 17, cex = 1.5)
points(df[1], te[1], col = "palegreen3", pch = 17, cex = 1.5)
points(df[which.min(te)], mse[which.min(te)], col = "darkorange",
    pch = 16, cex = 1.5)
points(df[which.min(te)], te[which.min(te)], col = "darkorange",
    pch = 16,cex = 1.5)
points(df[15], mse[15], col = "steelblue", pch = 15, cex = 1.5)
points(df[15], te[15], col = "steelblue", pch = 15, cex = 1.5)
legend(x = "top", legend = c("Training error", "Test error"),
lwd = rep(2, 2), col = c(gray(0.4), "orange3"), text.width = 0.3,
    cex = 0.8)
```

Figure 86 shows that the prediction error on the training dataset keeps decreasing with the increase of the `df`. This is consistent with our theory, and this only indicates a universal phenomenon that *a more complex model can fit the training data better*. On the other hand, we could observe that the testing error curve shows a **U-shaped** curve, indicating that an optimal model[128] exists in this range of the model complexity.

[128] I.e., the dip location on the U-shaped curve is where the optimal `df` could be found.

As this is an observation made on one dataset that was randomly generated, we should repeat this experiment multiple times to see if our observation is robust. The following R code repeats this experiment 100 times and presents the results in Figure 87.

```r
# Repeat the above experiments in 100 times
n_rep <- 100
n_train <- 50
coef <- c(-0.68,0.82,-0.417,0.32,-0.68)
v_noise <- 0.2
n_df <- 20
df <- 1:n_df
xy <- res <- list()
xy_test <- gen_data(n_test, coef, v_noise)
for (i in 1:n_rep) {
  xy[[i]] <- gen_data(n_train, coef, v_noise)
  x <- xy[[i]][, "x"]
  y <- xy[[i]][, "y"]
  res[[i]] <- apply(t(df), 2,
             function(degf) lm(y ~ ns(x, df = degf)))
}


# Compute the training and test errors for each model
pred <- list()
mse <- te <- matrix(NA, nrow = n_df, ncol = n_rep)
for (i in 1:n_rep) {
  mse[, i] <- sapply(res[[i]],
             function(obj) deviance(obj)/nobs(obj))
  pred[[i]] <- mapply(function(obj, degf) predict(obj,
                 data.frame(x = xy_test$x)),res[[i]], df)
  te[, i] <- sapply(as.list(data.frame(pred[[i]])),
             function(y_hat) mean((xy_test$y - y_hat)^2))
  }

# Compute the average training and test errors
av_mse <- rowMeans(mse)
av_te <- rowMeans(te)

# Plot the errors
plot(df, av_mse, type = "l", lwd = 2, col = gray(0.4),
     ylab = "Prediction error",
     xlab = "The degrees of freedom (logged) of the model",
     ylim = c(0.7*min(mse), 1.4*max(mse)), log = "x")
for (i in 1:n_rep) {
  lines(df, te[, i], col = "lightyellow2")
}
for (i in 1:n_rep) {
  lines(df, mse[, i], col = gray(0.8))
}
lines(df, av_mse, lwd = 2, col = gray(0.4))
lines(df, av_te, lwd = 2, col = "orange3")
points(df[1], av_mse[1], col = "palegreen3", pch = 17, cex = 1.5)
points(df[1], av_te[1], col = "palegreen3", pch = 17, cex = 1.5)
points(df[which.min(av_te)], av_mse[which.min(av_te)],
       col = "darkorange", pch = 16, cex = 1.5)
```

```
points(df[which.min(av_te)], av_te[which.min(av_te)],
        col = "darkorange", pch = 16, cex = 1.5)
points(df[20], av_mse[20], col = "steelblue", pch = 15, cex = 1.5)
points(df[20], av_te[20], col = "steelblue", pch = 15, cex = 1.5)
legend(x = "center", legend = c("Training error", "Test error"),
        lwd = rep(2, 2), col = c(gray(0.4), "darkred"),
        text.width = 0.3, cex = 0.85)
```

Again, we can see that the training error keeps decreasing when the model complexity increases, while the testing error curve has a U-shape. The key to identify the best model complexity is to locate the lowest point on the U-shaped error curve obtained from a testing dataset.

These experiments show that we need a testing dataset to evaluate the model to guide the model selection. Suppose you don't have a testing dataset. The essence of a testing dataset is that it is not used for training the model. So you create one. What the hold-out, random sampling, and cross-validation approaches really do is use the training dataset to generate an estimate of the error curve that is supposed to be obtained from a testing dataset[129].

To see that, let's consider a scenario that there are 200 samples, and a client has split them into two parts, i.e., a training dataset with 100 samples and a testing dataset with another 100 samples. The client only sent the training dataset to us. So we use the 10-fold cross-validation on the training dataset, using the following R code, to evaluate the models (df from 0 to 20).

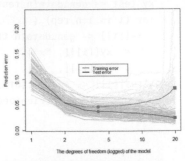

Figure 87: Prediction errors of the models (from df = 0 to df = 20) on the training dataset and testing dataset of 100 replications. The two highlighted curves represent the mean curves of the 100 replications of the training and testing error curves, respectively

[129] See Figure 94 and its associated text in the Remarks section for more discussion.

```
# Cross-validation
set.seed(seed)

n_train <- 100
xy <- gen_data(n_train, coef, v_noise)
x <- xy$x
y <- xy$y

fitted_models <- apply(t(df), 2,
        function(degf) lm(y ~ ns(x, df = degf)))
mse <- sapply(fitted_models,
        function(obj) deviance(obj)/nobs(obj))

n_test <- 100
xy_test <- gen_data(n_test, coef, v_noise)
pred <- mapply(function(obj, degf)
    predict(obj, data.frame(x = xy_test$x)),
    fitted_models, df)
te <- sapply(as.list(data.frame(pred)),
    function(y_hat) mean((xy_test$y - y_hat)^2))
```

```
n_folds <- 10
folds_i <- sample(rep(1:n_folds, length.out = n_train))
cv_tmp <- matrix(NA, nrow = n_folds, ncol = length(df))
for (k in 1:n_folds) {
  test_i <- which(folds_i == k)
  train_xy <- xy[-test_i, ]
  test_xy <- xy[test_i, ]
  x <- train_xy$x
  y <- train_xy$y
  fitted_models <- apply(t(df), 2, function(degf) lm(y ~
                               ns(x, df = degf)))
  x <- test_xy$x
  y <- test_xy$y
  pred <- mapply(function(obj, degf) predict(obj,
            data.frame(ns(x, df = degf))),
            fitted_models, df)
  cv_tmp[k, ] <- sapply(as.list(data.frame(pred)),
            function(y_hat) mean((y - y_hat)^2))
}
cv <- colMeans(cv_tmp)
```

Then we can visualize the result in Figure 88 (the R script is shown below). Note that, in Figure 88, we overlay the result of the 10-fold cross-validation (based on the 100 training samples) with the prediction error on the testing dataset to get an idea about how closely the 10-fold cross-validation can approximate the testing error curve[130].

```
# install.packages("Hmisc")
require(Hmisc)
plot(df, mse, type = "l", lwd = 2, col = gray(0.4),
     ylab = "Prediction error",
     xlab = "The degrees of freedom (logged) of the model",
     main = paste0(n_folds,"-fold Cross-Validation"),
     ylim = c(0.8*min(mse), 1.2*max(mse)), log = "x")
lines(df, te, lwd = 2, col = "orange3", lty = 2)
cv_sd <- apply(cv_tmp, 2, sd)/sqrt(n_folds)
errbar(df, cv, cv + cv_sd, cv - cv_sd, add = TRUE,
       col = "steelblue2", pch = 19,
lwd = 0.5)
lines(df, cv, lwd = 2, col = "steelblue2")
points(df, cv, col = "steelblue2", pch = 19)
legend(x = "topright",
       legend = c("Training error", "Test error",
                  "Cross-validation error"),
       lty = c(1, 2, 1), lwd = rep(2, 3),
       col = c(gray(0.4), "darkred", "steelblue2"),
       text.width = 0.4, cex = 0.85)
```

Overall, Figure 88 shows that the 10-fold cross-validation could generate fair evaluation of the models just like an independent and

[130] Remember that, in practice, we will not have access to the testing data, but we want our model to succeed on the testing data, i.e., to obtain the lowest error on the testing error curve. Thus, using cross-validation to mimic this testing procedure based on our training data is a "rehearsal".

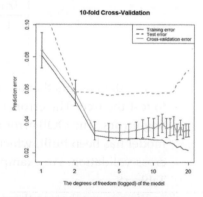

Figure 88: Prediction errors of the models (from `df` = 0 to `df` = 20) on the training dataset without cross-validation, on the training dataset using 10-fold cross-validation, and testing data of 100 samples, respectively

unseen testing dataset. Although its estimation of the error is smaller than the error estimation on the testing dataset, they both point towards the same range of model complexity that will neither overfit nor underfit the data.

Out-of-bag error in random forests

The random forest model provides a concept named out-of-bag error that plays a similar role as the hold-out method. Let's revisit how it works.

Rationale and formulation

Suppose that we have a training dataset of 5 instances (IDs as $1, 2, 3, 4, 5$). A random forest model with 3 trees is built. The 3 Bootstrapped datasets are shown in Table 16.

Tree	Bootstrap
1	$1, 1, 4, 4, 5$
2	$2, 3, 3, 4, 4$
3	$1, 2, 2, 5, 5$

Table 16: Three trees and the bootstrapped datasets

Since Bootstrap *randomly* selects samples from the original dataset to form Bootstrapped datasets, some data points in the original dataset may not show up in the Bootstrapped datasets. These data points are called **out-of-bag samples (OOB)** samples. For instance, for the random forest model that corresponds to Table 16, the OOB samples for each tree are shown in Table 17.

Tree	OOB samples
1	$2, 3$
2	$1, 5$
3	$3, 4$

Table 17: Out-of-bag (OOB) samples

The data points that are not used in training a tree could be used to test the tree. The errors on the OOB samples are called the **out-of-bag errors**. The OOB error can be calculated after a random forest model has been built, which seems to be computationally easier than cross-validation. An example to compute the OOB errors is shown in Table 18.

We can see that, as the data instance ($ID = 1$) is not used in training *Tree* 2, we can use *Tree* 2 to predict on this data instance, and we see that it correctly predicts the class as $C1$. Similarly, *Tree* 1 is used to predict on data instance ($ID = 2$), and the prediction is wrong. Overall, the OOB error of the random forest model is $1/6$.

Data ID	True label	Predicted label		
		Tree 1	*Tree 2*	*Tree 3*
1	C1		C1	
2	C2	C1		
3	C2	C2		C2
4	C1			C1
5	C2		C2	

Table 18: Out-of-bag (OOB) errors

Theory and method

The OOB error provides a computationally convenient approach to evaluate the random forest model without using a testing dataset or a cross-validation procedure. A technical concern is whether this idea can scale up. In other words, are there enough OOB samples to ensure that the OOB error is a fair and robust performance metric?

Recall that, for a random forest model with K trees, each tree is built on a Bootstrapped dataset from the original training dataset D. There are totally K Bootstrapped datasets, denoted as $B_1, B_2, \ldots, B_K$.

Usually, the size of each Bootstrapped dataset is the same size (denoted as N) as the training dataset D. Each data point in the Bootstrapped dataset is randomly and *independently* selected. Therefore, the probability of a data point from the training dataset D missing from a Bootstrapped dataset is[131]

$$\left(1 - \frac{1}{N}\right)^N.$$

[131] Because there are N independent trials of random selection, for a data point not to be selected, it has to be missed N times. And the probability for "not to be selected" is $\left(1 - \frac{1}{N}\right)$.

When N is sufficiently large, we have

$$\lim_{N \to \infty} \left(1 - \frac{1}{N}\right)^N = e^{-1} \approx 0.37.$$

Therefore, roughly 37% of the data points from D are not contained in a Bootstrapped dataset B_i, and thus, not used for training the *tree i*. These excluded data points are the OOB samples for the *tree i*.

As there are 37% of probability that a data point is not used for training a tree, we can infer that, on average, a data point is not used for training about 37% of the trees[132]. In other words, for each data point, in theory 37% of the trees are trained without this data point. This is a sizeable amount of data points, ensuring that the OOB error could be a stable and accurate evaluation of the model's performance on *future unseen testing data*.

[132] Note that the assumption is the Bootstrapped dataset has the same size as the original dataset, and the sampling is with replacement.

R Lab

We design a numeric study to compare the OOB error with the error obtained by a validation procedure and the error estimated on the training dataset. The three types of error rates are plotted in Figure 89.

First, we split the dataset into two halves: one for training and one for testing.

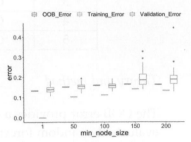

Figure 89: Comparison of different types of error rates

```
library(dplyr)
library(tidyr)
library(ggplot2)
require(randomForest)
set.seed(1)

library(RCurl)
url <- paste0("https://raw.githubusercontent.com",
              "/analyticsbook/book/main/data/AD.csv")
data <- read.csv(text = getURL(url))

target_indx <- which(colnames(data) == "DX_bl")
data[, target_indx] <- as.factor(paste0("c", data[, target_indx]))
rm_indx <- which(colnames(data) %in% c("ID", "TOTAL13", "MMSCORE"))
data <- data[, -rm_indx]

para.v <- c(1, 50, 100, 150, 200)
results <- NULL
```

Then, we build a set of random forest models by tuning the parameter `nodesize`, and obtain the OOB errors of the models.

```
# OOB error
for (ipara in para.v) {
rf <- randomForest(DX_bl ~ ., nodesize = ipara, data = data)
# nodesize = inodesize
results <- rbind(results, c("OOB_Error",
                  ipara, mean(rf$err.rate[, "OOB"])))
}
```

We also use the random sampling method to evaluate the errors of the models.

```
# Validation error
for (ipara in para.v) {
for (i in 1:50) {
train.ix <- sample(nrow(data), floor(nrow(data)/2))
rf <- randomForest(DX_bl ~ ., nodesize = ipara,
                  data = data[train.ix,
])
pred.test <- predict(rf, data[-train.ix, ], type = "class")
```

```
this.err <- length(
     which(pred.test != data[-train.ix, ]$DX_bl))/length(pred.test)
results <- rbind(results, c("Validation_Error", ipara, this.err))
}
}
```

Then, we obtain the training errors of the models.

```
# Training error
for (ipara in para.v) {
rf <- randomForest(DX_bl ~ ., nodesize = ipara, data = data)
# nodesize = inodesize
pred <- predict(rf, data, type = "class")
this.err <- length(which(pred != data$DX_bl))/length(pred)
results <- rbind(results, c("Training_Error", ipara, this.err))
}

colnames(results) <- c("type", "min_node_size", "error")
results <- as.data.frame(results)
results$error = as.numeric(as.character(results$error))
results$min_node_size <- factor(results$min_node_size,
                                unique(results$min_node_size))
ggplot() + geom_boxplot(data = results,
                        aes(y = error, x = min_node_size,
                            color = type)) +
           geom_point(size = 3)
```

Figure 89 shows that the OOB error rates are reasonably aligned with the testing error rates, while the training error rates are deceptively smaller.

The following R code conducts another numeric experiment to see if the number of trees impacts the OOB errors. In particular, we compare 50 trees with 500 trees, with their OOB errors plotted in Figure 90. On the other hand, we also observe that by increasing the number of trees, the OOB error decreases. This phenomenon is not universal (i.e., it is not always observed in all the datasets), but it does indicate the limitation of the OOB error: it is not as robust as the random sampling or cross-validation methods in preventing overfitting. But overall, the idea of OOB is inspiring.

```
para.v <- c(1, 50, 100, 150, 200)
results <- NULL

# OOB error with 500 trees
for (ipara in para.v) {
  rf <- randomForest(DX_bl ~ ., nodesize = ipara, ntree = 500,
                     data = data)
  # nodesize = inodesize
  results <- rbind(results, c("OOB_Error_500trees", ipara,
```

```
                              mean(rf$err.rate[,"OOB"])))
}

# OOB error with 50 trees
for (ipara in para.v) {
  rf <- randomForest(DX_bl ~ ., nodesize = ipara, ntree = 50,
                     data = data)  # nodesize = inodesize
  results <- rbind(results, c("OOB_Error_50trees", ipara,
                     mean(rf$err.rate[,"OOB"])))
}
colnames(results) <- c("type", "min_node_size", "error")
results <- as.data.frame(results)
results$error = as.numeric(as.character(results$error))
results$min_node_size <- factor(results$min_node_size,
                           unique(results$min_node_size))
ggplot() + geom_boxplot(data = results,
                 aes(y = error, x = min_node_size,
                     fill = type)) +
          geom_bar(stat = "identity",position = "dodge")
```

Remarks

The "law" of learning errors

We have seen the *R-squared* could be manipulated to become larger, i.e., by adding into the model with more variables even if these variables are not predictive. This *bug* is not a special trait of the linear regression model only. The *R-squared* by its definition is computed based on the training data, and therefore, is essentially a *training error*. For any model that offers a flexible degree of complexity (e.g., examples are shown in Table 19), its *training error* could be decreased if we make the model more complex.

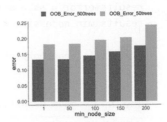

Figure 90: OOB error rates from random forests with a different number of trees

Model	Complexity parameter
Linear regression	Number of variables
Decision tree	Depth of tree
Random forest	Number of trees
	Depth of trees
	Number of variables to be selected for each split

Table 19: The complexity parameters of some models

For example, let's revisit the decision tree model shown in Figure 47 in **Chapter 3**. A deeper tree segments the space into smaller rectangular regions, guided by the distribution of the *training data*, as shown in Figure 91. The model achieves 100% accuracy—but this is an illusion, since the training data contains noise that could not be predicted. These rectangular regions, particularly those smaller ones,

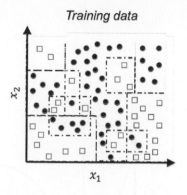
Training data

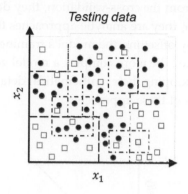
Testing data

Figure 91: A much more complex decision tree model than the one in Figure 47; (left) the tree model perfectly fits the *training data*; (right) the tree performs poorly on the *testing data*

are susceptible to the noise. When we apply this deeper tree model on a *testing data* that is sampled from the same distribution of the *training data*[133], the model performs poorly.

It is generally true that the more complex a model gets, the lower the error on the training dataset becomes, as shown in Figure 92 (left). This is the "law" of the *training error*, and training a model based on the training error could easily "spoil" the model. If there is a testing dataset, the error curve would look like *U-shaped*, as shown in Figure 92 (middle), and the curve's dip point helps us identify the best model complexity. While on the other hand, if there is no testing dataset, we could use cross-validation to obtain error estimates. The error curve obtained by cross-validation on the training data, as shown in Figure 92 (right), should provide a good approximation of the error curve of the testing data. The three figures in Figure 92, from left to right, illustrate a big picture of the *laws* of the errors and why some techniques such as the cross-validation have central importance in data analytics.

[133] The overall *morphology* of the two datasets looks alike; the differences, however, are due to the noise that is unpredictable.

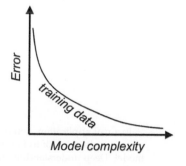
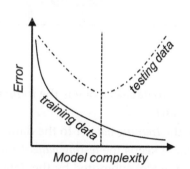
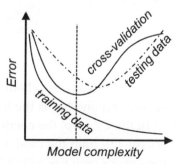

Figure 92: The "law" of learning errors

There are other approaches that play similar roles as the cross-validation, i.e., to approximate the error curve on unseen testing data. Examples include the **Akaike information criterion (AIC)**, the **Bayesian information criterion (BIC)**, and many other model selection criteria alike. Different from the cross-validation, they don't resample the training data. Rather, they are analytic approaches that evaluate a model's performance by offsetting the model's training error with a complexity penalty, i.e., the more complex a model gets, the larger the penalty imposed. Skipping their mathematical details, Figure 93 illustrates the basic idea of these approaches.

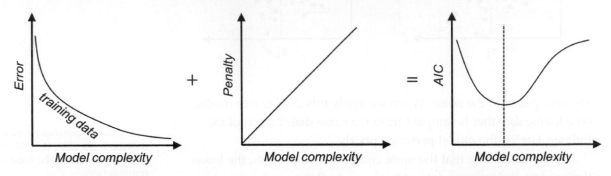

Figure 93: The basic idea of the AIC and BIC criteria

A larger view of model selection and validation

The practice of data analytics has evolved and developed an elaborate process to protect us from overfitting or underfitting a model. The 5-step process is illustrated in Figure 94.

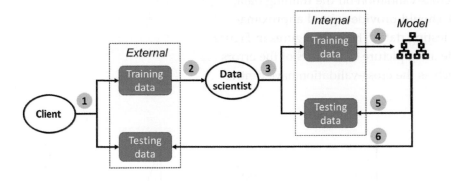

Figure 94: A typical process of how data scientists work with clients to develop robust models

In the 1st step, the client collects two datasets, one is the *training dataset* and another is the *testing dataset*.

In the 2nd step, the client sends the *training dataset* to the data scientist to train the model. The client keeps the *testing dataset* for the client's own use to test the final model submitted by the data scientist.

Now the data scientist should keep in mind that, no matter how the model is obtained[134], its goal is to predict well on the *unseen testing dataset*. How shall we do so, without access to the *testing dataset*?

Just like in Bootstrap, we mimic the process.

[134] In a real application, you may try all you could think of to find your best model. Deep understanding of your models always help. Sometimes it is also luck, insight, and hard-working trial and error. What matters is your model is really good and can outperform your competitor's. Data scientists survive in a harsh competitive environment.

In the 3rd step, the data scientist mimics the testing procedure as the client would use. The data scientist splits the *training dataset* into two parts, one for model training and one for model testing[135].

In the 4th step, the data scientist creates a model that should fit the *internal training dataset* well. Cross-validation is often used in this step.

In the 5th step, the data scientist tests the model obtained in the 4th step using the *internal testing data*. This is the final pass that will be conducted in house, before the final model is submitted to the client. Note that, the 5th step could not be integrated into the model selection process conducted in the 4th step—otherwise, the *internal testing data* is essentially used as an *internal training dataset*[136].

In the 6th step, the data scientist submits the final model to the client. The model will be evaluated by the client on the *internal testing dataset*. The data scientist may or may not learn the evaluation result of the final model from the client.

The confusion matrix

The *rare disease* example mentioned earlier in this chapter implies that the context matters. It also implies that *how we evaluate a model's performance* matters as well.

Accuracy, naturally, is a most important evaluation metric. As any *overall* evaluation metric, it averages things and blurs boundaries between categories, and for the same reason, it could be broken down into more *sub*categories. For example, a binary classification problem has two classes. We often care about specific accuracy on either class, i.e., if one class represents disease (positive) while another represents normal (negative), as a convention in medicine, we name the correct prediction on a positive case as **true positive** (**TP**) and name the correct prediction on a negative case as **true negative** (**TN**). Correspondingly, we define the **false positive** (**FP**) as incorrect prediction on a true negative case, and **false negative** (**FN**) as incorrect prediction on a true positive case. This is illustrated in Table 20, the so-called **confusion matrix**.

The confusion matrix		Reality	
		Positive	*Negative*
Prediction	*Positive*	True positive (**TP**)	False positive (**FP**)
	Negative	False negative (**FN**)	True negative (**TN**)

Based on *TP*, the concept **true positive rate** (**TPR**) could also be defined, i.e., *TPR* = TP/(TP+FN). Similarly, we can also define the

[135] Generate a "training dataset" and a "testing dataset" from the *training dataset*. To avoid confusion, these two are often called **internal** training dataset and **internal** testing dataset, respectively. The training and testing datasets the client creates are often called **external** training dataset and **external** testing dataset, respectively.

[136] After all, the usage of the dataset dictates its name.

Table 20: The confusion matrix

false positive rate (FPR) as FPR = FP/(FP+TN).

The ROC curve

Building on the *confusion matrix*, the **receiver operating characteristic curve (ROC curve)** is an important evaluation metric for classification models.

Recall that, in a logistic regression model, before we make the final prediction, an intermediate result is obtained first

$$p(x) = \frac{1}{1 + e^{-\left(\beta_0 + \Sigma_{i=1}^{p} \beta_i x_i\right)}}.$$

A **cut-off value**[137] is used to make the binary predictions, i.e., it classifies the cases whose $p(x)$ are larger than the cut-off value as *positive*; otherwise, if $p(x)$ is smaller than the cut-off value, *negative*. This means that, for each cut-off value, we can obtain a confusion matrix with different values of the TP, FP, FN, and TN. As there are many possible cut-off values, the *ROC curve* is a succinct way to synthesize all the scenarios of all possible cut-off values, i.e., it tries many cut-off values and plots the FPR (x-axis) against the TPR (y-axis). This is illustrated in Figure 95.

[137] By default, 0.5.

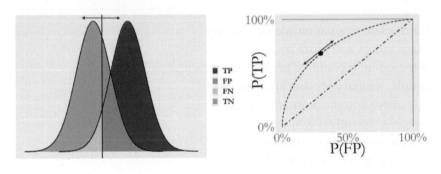

Figure 95: The logistic model produces an intermediate result $p(x)$ for the cases of both classes: (left) shows the distributions of $p(x)$ of both classes and a particular cut-off value; and (right) shows the ROC curve that synthesizes all the scenarios of all the cut-off values

The ROC curve is more useful to evaluate a model's *potential*, i.e., Figure 95 presents the performances of the logistic regression model for *all* cut-off values rather than *one* cut-off value. The 45° line represents a model that is equivalent to *random guess*. In other words, the ROC curve of a model that lacks potential for prediction will be close to the 45° line. A better model will show a ROC curve that is closer to the upper left corner point. Because of this, the **area under the curve (AUC)** is often used to summarize the ROC curve of a model. The higher the AUC, the better the model.

A Small Data Example   Let's study how a ROC curve could be created using an example. Consider a random forest model of 100 trees and its prediction on 9 data points. A random forest model uses the *majority voting* to aggregate the predictions of its trees to reach a final binary prediction. The *cut-off value* concerned here is the threshold of votes, i.e., here, we try three cut-off values, C=50 (default in `randomForest`), C=37, and C=33, as shown in Table 21.

ID	Vote	True label	Predicted label		
			C=50	C=37	C=33
1	38	1	0	1	1
2	49	1	0	1	1
3	48	0	0	1	1
4	76	1	1	1	1
5	32	0	0	0	0
6	57	0	1	1	1
7	36	1	0	0	1
8	36	0	0	0	1
9	35	0	0	0	1

Table 21: Prediction on 9 data points via a random forest model of 100 trees, with different cut-off values of the vote threshold, C=50 (default in `randomForest`), C=37, and C=33

Based on the definition of the confusion matrix in Table 20, we calculate the metrics in Table 22.

	C=50	C=37	C=33
Accuracy	5/9	6/9	5/9
TP	1	3	4
FP	1	2	4
FN	3	1	0
TN	4	3	1
FPR = FP/(FP+TN)	$1/(1+4)$	$2/(2+3)$	$4/(4+1)$
TPR = TP/(TP+FN)	$1/(1+3)$	$3/(3+1)$	$4/(4+0)$

Table 22: Metrics for predictions in Table 21

With three cut-off values, we map the three points in Figure 96 by plotting the *FPR* (x-axis) against the *TPR* (y-axis). There are a few R packages to generate a ROC curve for a classification model. Figure 96 illustrates the basic idea implemented in these packages to draw a ROC curve: sample a few cut-off values and map a few points in the figure, then draw a smooth curve that connects the point.

R Example   We build a logistic regression model using the AD data as we have done in **Chapter 3**.

```
# ROC and more performance metrics of logistic regression model
# Load the AD dataset
library(RCurl)
url <- paste0("https://raw.githubusercontent.com",
              "/analyticsbook/book/main/data/AD.csv")
AD <- read.csv(text=getURL(url))
str(AD)
```

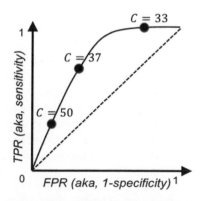

Figure 96: Illustration of how to draw a ROC curve using the data in Tables 21 and 22

```
# Split the data into training and testing sets
n = dim(AD)[1]
n.train <- floor(0.8 * n)
idx.train <- sample(n, n.train)
AD.train <- AD[idx.train,]
AD.test <- AD[-idx.train,]

# Automatic selection of the model
logit.AD.full <- glm(DX_bl ~ ., data = AD.train[,c(1:16)],
                     family = "binomial")
logit.AD.final <- step(logit.AD.full, direction="both", trace = 0)
summary(logit.AD.final)
```

Then we use the function, `confusionMatrix()` from the R package `caret` to obtain the confusion matrix

```
require(e1071)
require(caret)
# Prediction scores
pred = predict(logit.AD.final, newdata=AD.test,type="response")
confusionMatrix(data=factor(pred>0.5), factor(AD.test[,1]==1))
```

The result is shown below.

```
## Confusion Matrix and Statistics
##
##           Reference
## Prediction FALSE TRUE
##      FALSE    48    7
##      TRUE      7   42
##
##                Accuracy : 0.8654
##                  95% CI : (0.7845, 0.9244)
##     No Information Rate : 0.5288
##     P-Value [Acc > NIR] : 3.201e-13
##
##                   Kappa : 0.7299
##  Mcnemar's Test P-Value : 1
##
##             Sensitivity : 0.8727
##             Specificity : 0.8571
##          Pos Pred Value : 0.8727
##          Neg Pred Value : 0.8571
##              Prevalence : 0.5288
##          Detection Rate : 0.4615
##    Detection Prevalence : 0.5288
##       Balanced Accuracy : 0.8649
##
##        'Positive' Class : FALSE
##
```

The ROC curve could be drawn using the R Package `ROCR`.

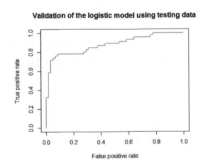

Figure 97: ROC curve of the logistic regression model

```
# Generate the ROC curve using the testing data
# Compute ROC and Precision-Recall curves
require('ROCR')
linear.roc.curve <- performance(prediction(pred, AD.test[,1]),
                         measure='tpr', x.measure='fpr' )
plot(linear.roc.curve,  lwd = 2, col = "orange3",
  main = "Validation of the logistic model using testing data")
```

The ROC curve is shown in Figure 97.

Exercises

1. A random forest model is built on the training data with 6 data
   points. The details of the trees and their bootstrapped datasets are
   shown in Table 23.

Bootstrapped data	Tree
1,3,4,4,5,6	1
2,2,4,4,4,5	2
1,2,2,5,6,6	3
3,3,3,4,5,6	4

Table 23: Bootstrapped datasets and the built trees

To calculate the out-of-bag (OOB) errors, which legitimate data
points are to be used for each tree? You can mark them out in
Table 24.

Tree	Bootstrapped data	1(C1)	2(C2)	3(C2)	4(C1)	5(C2)	6(C1)
1	1,3,4,4,5,6						
2	2,2,4,4,4,5						
3	1,2,2,5,6,6						
4	3,3,3,4,5,6						

Table 24: Mark the elements where OOB errors could be collected

2. Figure 98 shows the ROC curves of two classification models.
   Which model is better?

3. Follow up on the simulation experiment in Q9 in **Chapter 2** and
   the random forest model in Q5 in **Chapter 4**. Split the data into
   a training set and a testing test, then use 10-fold cross-validation
   to evaluate the performance of the random forest model with 100
   trees.

4. Follow up on Q3. Increase the sample size of the experiment to
   1000, and comment on the result.

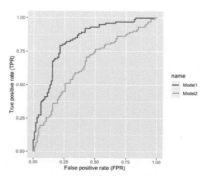

Figure 98: The ROC curve of two models

Chapter 6: Diagnosis
Residuals & Heterogeneity

Overview

Chapter 6 is about *Diagnosis*. Diagnosis, in one sense, is to see if the assumptions of the model match the empirical characteristics of the data. For example, the t-test of linear regression model builds on the normality assumption of the errors. If this assumption is not met by the data, the result of the t-test is concerned. Departure from assumptions doesn't always mean that the model is not useful[138]. The gap between the theoretical assumptions of the model and the empirical data characteristics, together with the model itself, should be taken as a whole when we evaluate the strength of the conclusion. This wholesome idea is what diagnosis is about. It also helps us to identify opportunities to improve the model. Models are representations/approximations of reality, so we have to be critical about them, yet being critical is different from being dismissive[139]. There are many diagnostic tools that we can use to strengthen our critical evaluation.

Many diagnostic tools focus on the **residual analysis**. Residuals provide a numerical evaluation of the difference between the model and the data. Recall that y denotes the observed value of the outcome variable, $f(x)$ denotes the model, and $\hat{y}$ denotes the prediction (i.e., $\hat{y} = f(x)$ is the prediction made by the model on the data point x). The residual, denoted as $\hat{e}$, is defined as $\hat{e} = \hat{y} - y$. For any model that is trained on N data points, we could obtain N residuals, and draw the residuals as shown in Figure 99:

1. Figure 99 (left). There is a linear relationship between $\hat{e}$ and $\hat{y}$, which suggests an absurd fact: $\hat{y}$ could be used as a predictor to predict the $\hat{e}$, the *error*. For instance, when $\hat{y} = -1$, the error is between 1 and 2. If we adjust the prediction to be $\hat{y} + 1$, wouldn't that make the error to be between 0 and 1? A reduced error means a better prediction model.

[138] *"All models are wrong, some are useful."*— George Box.

[139] A model that doesn't fit the data also generates knowledge—revealed not by the failed model but by the fact that this model actually misfits. See Jaynes, E.T., *Probability Theory: the Logic of Science.* Cambridge Press, 2003.

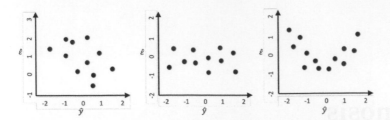

Figure 99: Suppose that three models are built on a dataset, and their residual plots are drawn: (left) decision tree; (middle) RF; (right) linear regression

To generalize this, let's build another model $g[f(x)]$ that takes $f(x)$ as the predictor to predict $\hat{e}$. Then, we can combine the two models, $f(x)$ and $g[f(x)]$, and obtain an improved prediction $\hat{y}$ as $f(x) + g[f(x)]$. This is shown in Figure 100.

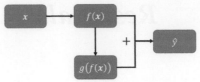

Figure 100: A new model, inspired by the pattern seen in Figure 99 (left)

2. Figure 99 (middle). No correlation between $\hat{e}$ and $\hat{y}$ is observed. In other words, knowing $\hat{y}$ offers no help to predict $\hat{e}$. This is what a good model would behave like.

3. Figure 99 (right). There is a piece-wise linear relationship between $\hat{e}$ and $\hat{y}$. If we segment the figure by a vertical line at zero, we could apply the same argument made in Figure 99 (left) for each piece here: the model could be further improved following the same strategy outlined in Figure 100.

As each data point contributes a residual, the *residual analysis* offers us opportunities to examine some collective phenomena to improve the overall quality of the model. It also helps us check local patterns where we may find areas of improvement of the model or particularities of the data that the model could not synthesize. The beauty of checking out the residuals is that there is always something that is beyond our experience and expectation.

Diagnosis in regression

Residual analysis

The R package `ggfortify` provides a nice bundle that includes the **residual analysis, cook's distance, leverage**, and **Q-Q plot**.

Let's use the final regression model we identified in **Chapter 2** for an example.

```
library(RCurl)
url <- paste0("https://raw.githubusercontent.com",
              "/analyticsbook/book/main/data/AD.csv")
AD <- read.csv(text=getURL(url))
AD$ID = c(1:dim(AD)[1])
str(AD)
# fit a full-scale model
```

```
AD_full <- AD[,c(2:17)]
lm.AD <- lm(MMSCORE ~ ., data = AD_full)
summary(lm.AD)
# Automatic model selection
lm.AD.F <- step(lm.AD, direction="backward", test="F")
```

Details of the model are shown below.

```
## MMSCORE ~ PTEDUCAT + FDG + AV45 + HippoNV + rs744373 + rs610932
##       + rs3764650 + rs3865444
##
##              Df Sum of Sq     RSS    AIC F value     Pr(>F)
## <none>                    1537.5 581.47
## - rs3764650   1     7.513 1545.0 581.99  2.4824   0.115750
## - rs744373    1    12.119 1549.6 583.53  4.0040   0.045924 *
## - rs610932    1    14.052 1551.6 584.17  4.6429   0.031652 *
## - rs3865444   1    21.371 1558.9 586.61  7.0612   0.008125 **
## - AV45        1    50.118 1587.6 596.05 16.5591 5.467e-05 ***
## - PTEDUCAT    1    82.478 1620.0 606.49 27.2507 2.610e-07 ***
## - HippoNV     1   118.599 1656.1 617.89 39.1854 8.206e-10 ***
## - FDG         1   143.852 1681.4 625.71 47.5288 1.614e-11 ***
## ---
## Signif. codes:  0 '***' 0.001 '**' 0.01 '*' 0.05 '.' 0.1 ' ' 1
```

We use the `ggfortify` to produce 6 diagnostic figures as shown in Figure 101.

```
# Conduct diagnostics of the model
# install.packages("ggfortify")
library("ggfortify")
autoplot(lm.AD.F, which = 1:6, ncol = 3, label.size = 3)
```

The following is what we observe from Figure 101.

1. Figure 101 (upper left). This is the scatterplot of the residuals versus fitted values of the outcome variable. As we have discussed in Figure 99, this scatterplot is supposed to show purely random distributions of the dots. Here, we notice two abnormalities: (1) there is a relationship between the residuals and fitted values; and (2) there are unusual parallel lines[140]. These abnormalities have a few implications: (1) the linear model *underfits* the data, so a nonlinear model is needed; (2) we have assumed that the data points are independent with each other, now this assumption needs to be checked; and (3) we have assumed *homoscedasticity*[141] of the variance of the errors. This is another assumption that needs to be checked[142].

2. Figure 101 (upper right). The **Q-Q plot** checks the normality assumption of the errors. The 45° line is a fixed *baseline*, while the

[140] This is often observed if the outcome variable takes integer values.

[141] In **Chapter 2**, we assume that $\epsilon \sim N(0, \sigma_\varepsilon^2)$. It assumes the errors have the same variance, σ_ε^2, for all data points.

[142] To build nonlinear regression model or conditional variance regression model, see **Chapter 9**.

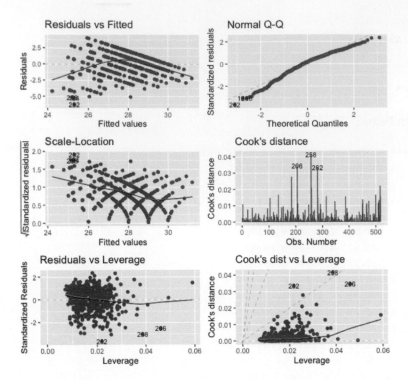

Figure 101: Diagnostic figures of regression model on the AD dataset

dots correspond to the data points. If the normality assumption is met, the dots should align with the line. Here, we see mild departure of the data from the normality assumption. And some particular data points such as the data points 282 and 256 are labelled since they are outstanding[143].

3. Figure 101 (middle left). This is a transformation of Figure 101 (upper left). Diagnostic tools are usually *opportunistic* approaches, i.e., what you see is what you get; if nothing particular is observed, it doesn't mean there is no anomaly in the data. Changing perspectives is a common practice in model diagnosis.

4. Figure 101 (middle right). The **Cook's distance** identifies influential data points that have *larger than average* influence on the parameter estimation of the model. For a data point x_i, its Cook's distance D_i is defined as the sum of all the changes in the regression model when x_i is removed from the training data. There is a closed-form formula[144] to compute D_i, for $j = 1, 2, \ldots, p$, based on the *least squares estimator* of the regression parameters.

5. Figure 101 (lower left). The **leverage** of a data point, on the other hand, shows the influence of the data point in another way. The leverage of a data point is defined as $\frac{\partial \hat{y}_i}{\partial y_i}$. This reflects how sensitively the prediction $\hat{y}_i$ is influenced by y_i. What data point will have a larger leverage value? For those surrounded by many

[143] Are those points outliers? The Q-Q plot provides no conclusive evidence. It only suggests.

[144] Cook, R.D., *Influential Observations in Linear Regression*, Journal of the American Statistical Association, Volume 74, Number 365, Pages 169-174, 1979.

closeby data points, their leverages won't be large: the impact of a data point's removal in a dense neighborhood is limited, given many other similar data points nearby. It is the data points in sparsely occupied neighborhoods that have large leverages. These data points could either be outliers that severely deviate from the linear trend represented by the majority of the data points, or could be valuable data points that align with the linear trend but lack neighbor data points. Thus, a data point that is influential doesn't necessarily imply it is an outlier, as shown in Figure 102. When a data point has a larger leverage value, in-depth examination of the data point is needed to determine which case it is.

6. Figure 101 (lower right). This is another form of showing the information that is presented in Figure 101 (middle right) and (lower left).

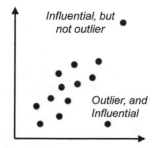

Figure 102: Outliers v.s. influential data points

A Simulation Experiment   We simulate a dataset while all the assumptions of the linear regression model are met. The model is

$$y = \beta_0 + \beta_1 x_1 + \beta_2 x_2 + \varepsilon, \varepsilon \sim N(0,1).$$

We simulate 100 samples from this model.

```
# For comparison, let's simulate data
# from a model that fits the assumptions
x1 <- rnorm(100, 0, 1)
x2 <- rnorm(100, 0, 1)
beta1 <- 1
beta2 <- 1
mu <- beta1 * x1 + beta2 * x2
y <- rnorm(100, mu, 1)
```

We fit the data using linear regression model.

```
lm.XY <- lm(y ~ ., data = data.frame(y,x1,x2))
summary(lm.XY)
```

The fitted model fairly reflects the underlying model.

```
##
## Call:
## lm(formula = y ~ ., data = data.frame(y, x1, x2))
##
## Residuals:
##     Min      1Q  Median      3Q     Max
## -2.6475 -0.6630 -0.1171  0.7986  2.5074
##
## Coefficients:
##              Estimate Std. Error t value Pr(>|t|)
```

```
## (Intercept)    0.0366    0.1089   0.336     0.738
## x1             0.9923    0.1124   8.825 4.60e-14 ***
## x2             0.9284    0.1159   8.011 2.55e-12 ***
## ---
## Signif. codes:  0 '***' 0.001 '**' 0.01 '*' 0.05 '.' 0.1 ' ' 1
##
## Residual standard error: 1.088 on 97 degrees of freedom
## Multiple R-squared:  0.6225, Adjusted R-squared:  0.6147
## F-statistic: 79.98 on 2 and 97 DF,  p-value: < 2.2e-16
```

```
autoplot(lm.XY, which = 1:6, ncol = 2, label.size = 3)
```

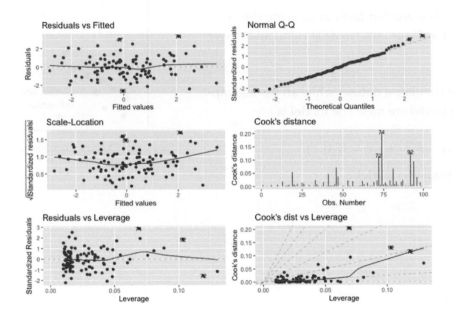

Figure 103: Diagnostic figures of regression model on a simulation dataset

Then, we generate the diagnostic figures as shown in Figure 103. Now Figure 103 provides a contrast of Figure 101. For example, in Figure 103 (upper left), we don't see a nonrandom statistical pattern. The relationship between the residual and fitted values seems to be null. From the *QQ-plot*, we see that the normality assumption is held well. On the other hand, from the *Cook's distance* and the *leverage*, some data points are observed to be outstanding, which are labeled. As we simulated the data following the assumptions of the linear regression model, this experiment shows that it is normal to expect a few data points to show outstanding *Cook's distance* and *leverage* values.

```
# Conduct diagnostics of the model
library("ggfortify")
autoplot(lm.XY, which = 1:6, ncol = 3, label.size = 3)
```

Multicollinearity

Multicollinearity refers to the phenomenon that there is *a high correlation among the predictor variables*. This causes a serious problem for linear regression models. We can do a simple analysis. Consider a linear system shown below

$$y = \beta_0 + \beta_1 x_1 + \beta_2 x_2 + \cdots + \beta_p x_p + \varepsilon_y,$$
$$\varepsilon_y \sim N\left(0, \sigma_{\varepsilon_y}^2\right).$$

This looks like a regular linear regression model. However, here we further have

$$x_1 = 2x_2 + \epsilon_x,$$
$$\epsilon_x \sim N\left(0, \sigma_{\epsilon_x}^2\right).$$

This *data-generating mechanism* is shown in Figure 104. It is a system that suffers from *multicollinearity*, i.e., if we apply a linear regression model on this system, the following models are *both* true models

$$y = \beta_0 + (2\beta_1 + \beta_2) x_2 + \beta_3 x_3 \ldots + \beta_p x_p$$
$$y = \beta_0 + (\beta_1 + 0.5\beta_2) x_1 + \beta_3 x_3 + \cdots + \beta_p x_p$$

The problem of multicollinearity results from an inherent ambiguity of the models that could be taken as faithful representation of the *data-generating mechanism*. If the *true* model is ambiguous, it is expected that an estimated model suffers from this problem as well.

There are some methods that we can use to diagnose *multicollinearity*. For instance, we may visualize the correlations among the predictor variables using the R package `corrplot`.

```
# Extract the covariance matrix of the regression parameters
Sigma = vcov(lm.AD.F)
# Visualize the correlation matrix of the estimated regression
# parameters
# install.packages("corrplot")
library(corrplot)
corrplot(cov2cor(Sigma), method="ellipse")
```

Figure 105 shows that there are significant correlations between the variables, `FDG` , `AV45` , and `HippoNV` , indicating a concern for multicollinearity. On the other hand, it seems that the correlations are moderate, and not all the variables are strongly correlated with each other.

It is of interest to see why the strong correlations among predictor variables cause problems in the *least squares* estimator of the regression coefficients. Recall that $\widehat{\beta} = \left(X^T X\right)^{-1} X^T y$. If there are strong correlations among predictor variables, the matrix $X^T X$ is

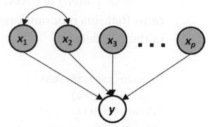

Figure 104: The *data-generating mechanism* of a system that suffers from *multicollinearity*

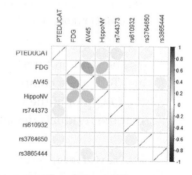

Figure 105: Correlations of the predictors in the regression model of `MMSCORE`

ill-conditioned, i.e., small changes on X result in large and unpredictable changes on the inverse matrix $X^T X$, which further causes instability of the parameter estimation in $\widehat{\beta}$.[145]

[145] To overcome multicollinearity in linear regression, the *Principal Component Analysis* discussed in **Chapter 8** is useful.

Diagnosis in random forests

Residual analysis

We can use the `plotmo` package to perform residual analysis for a random forest model. For instance, we build a random forest model to predict the variable `AGE` in the AD dataset. We plot the residual versus the fitted values as shown in Figure 106 which shows there is a linear pattern between the fitted values and residuals. This indicates that this random forest model missed some linear relationship in the AD dataset.

```
require(randomForest)
require(plotmo)
library(RCurl)
set.seed(1)
url <- paste0("https://raw.githubusercontent.com",
              "/analyticsbook/book/main/data/AD_hd.csv")
data <- read.csv(text=getURL(url))

target <- data$AGE
rm_indx <- which(colnames(data) %in%
                 c("AGE", "ID", "TOTAL13", "MMSCORE","DX_bl"))
X <- data[, -rm_indx]
rf.mod <- randomForest(X, target)
plotres(rf.mod, which = 3)
```

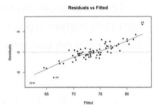

Figure 106: Residuals versus fitted in the random forest model

The random forest model doesn't assume normality of its residuals. To make a comparison with the linear regression model, we draw the Q-Q plot of the random forest model in Figure 107. It can be seen that the residuals deviate from the straight line.

```
plotres(rf.mod, which = 4)
```

As the random forest model is an algorithmic modeling approach that imposes no analytic assumption, diagnosis could still be done but interpretations are not as strong as in a linear regression model. There is still value to do so, to find area of improvement of the model, e.g., as Figure 106 suggests the random forest model could be further improved to incorporate the linear pattern in the data.

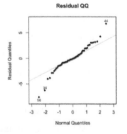

Figure 107: The Q-Q plot of residuals of the random forest model

Clustering

Rationale and formulation

Clustering takes the idea of *diagnosis* to a different level. If the *residual analysis* is like a tailor working out the perfect outfit for a client, clustering is ... well, it is better to see Figure 108 first.

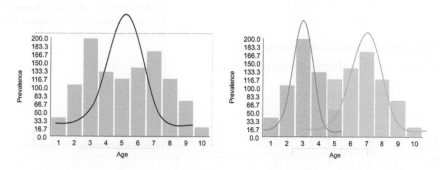

Figure 108: A "tailor" tries to (left) make an outfit (i.e., the normal curve) for a "client" (i.e., the data, represented as a histogram) vs. (right) then the "tailor" realizes the form of the outfit should be two normal curves

Figure 108 demonstrates one meaning of clustering: a dataset is heterogeneous and is probably collected from a few different populations (sometimes we call them *sub*populations). Understanding the clustering structure of a dataset not only benefits the statistical modeling, as shown in Figure 108 where we will use two normal distributions to model the data, but also reveals insights about the problem under study. For example, the dataset shown in Figure 108 was collected from a disease study of young children. It suggests that there are two disease mechanisms (we often call them two *phenotypes*). Phenotypes discovery is important for disease treatment, since patients with different disease mechanisms respond to treatments differently. A typical approach for phenotypes discovery is to collect an abundance of data from many patients. Then, we employ a range of algorithms to discover clusters of the data points. These clustering algorithms, differ from each other in their premises of what a cluster looks like, more or less bear the same conceptual framework as shown in Figure 108.

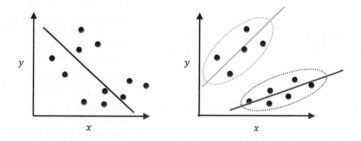

Figure 109: Another example of clustering: if the clustering structure is ignored, the fitted model (left) may show the opposite direction of the true model (right)

Clustering is a flexible concept that could be applied in other scenarios as well. Figure 109 demonstrates another meaning of clustering. It is less commonly perceived, but in practice it is not uncommon. The "moral of the story" shown in Figure 109 tells us that, when you have a dataset, you may want to conduct EDA and check the clustering structure first before imposing a model that may only fit the *data format* but not the *statistical structure*[146].

[146] E.g., in Figure 109: *data format*: we have predictors and outcome, so it seems natural to fit a linear regression model; *statistical structure*: however, it is a mix of two subpopulations that demand two models.

Theory and method

Given a dataset, how do we know there is a clustering structure? Consider the dataset shown in Table 25. Are there *sub*populations as shown in Figure 108?

	x_1	x_2	x_3	x_4	x_5	x_6	x_7
Value	1.13	4.76	0.87	3.32	4.29	1.03	0.98
Cluster	?	?	?	?	?	?	?
$\Theta: \pi_1 =?, \pi_2 =?; \mu_1 =?, \sigma_1^2 =?; \mu_2 =?, \sigma_2^2 =?$							

Table 25: Example of a dataset

A visual check of the 7 data points suggests there are probably two clusters. If each cluster can be modeled as a Gaussian distribution, this would be a two-component **Gaussian Mixture Model (GMM)**[147].

In this particular dataset, clustering could be done by learning the parameters of the two-component (**GMM**), (i.e., to address the question marks in the last row of Table 25). If we have known the parameters Θ, we could probabilistically infer which cluster each data point belongs to (i.e., to address the question marks in the second row of Table 25). On the other hand, if we have known which cluster each data point belongs to, we can collect the data points of each cluster to estimate the parameters of the Gaussian distribution that characterizes each cluster. This "locked" relation between the two tasks is shown in Figure 110.

The two interdependent tasks hold the key for each other. What is needed is *initialization*. As there are two blocks in Figure 110, we have two locations to initialize the process of unlocking.

[147] A GMM consists of multiple Gaussian distributions. Figure 108 shows one example of two univariate Gaussian distributions mixed together. Generally, the parameters of a GMM are denoted as Θ, which include the parameters of each Gaussian distribution: μ_i and σ_i are the mean and variance of the ith Gaussian distribution, respectively, and π_i is the proportion of the data points that were sampled from the ith Gaussian distribution.

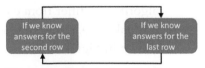

Figure 110: The locked relation between parameter estimation (M-step, i.e., last row of Table 25) and data point inference (E-step, i.e., second row of Table 25) in GMM

Initialization   Let's initialize the values in the second row of Table 25 for an example. We assign (i.e., *randomly*) labels on the data points as shown in Table 26.

M-step   Then, we estimate $\mu_1 = 1.75$ and $\sigma_1^2 = 2.83$ based on the data points $\{1.13, 4.76, 0.87, 1.03, 0.98\}$.[148]

Similarly, we could estimate $\mu_2 = 3.81$ and $\sigma_2^2 = 0.47$ based on the data points $\{3.32, 4.29\}$.

[148] These 5 data instances are initially assigned to $C1$. Note that 4.76 is different from the rest of the data points in the same cluster. This is an error introduced by the initialization. Later we will see that this error could be automatically fixed by the algorithm.

x_i	x_1	x_2	x_3	x_4	x_5	x_6	x_7
ID	1.13	4.76	0.87	3.32	4.29	1.03	0.98
Label	C1	C1	C1	C2	C2	C1	C1
Θ : $\pi_1 =?, \pi_2 =?; \mu_1 =?, \sigma_1^2 =?; \mu_2 =?, \sigma_2^2 =?$							

Table 26: Initialization on the dataset example

It is straightforward to estimate $\pi_1 = 5/7 = 0.714$ and $\pi_2 = 2/7 = 0.286$.

Table 26 is updated.

x_i	x_1	x_2	x_3	x_4	x_5	x_6	x_7
ID	1.13	4.76	0.87	3.32	4.29	1.03	0.98
Label	C1	C1	C1	C2	C2	C1	C1
Θ : $\pi_1 = 0.714, \pi_2 = 0.286; \mu_1 = 1.75, \sigma_1^2 = 2.83; \mu_2 = 3.81, \sigma_2^2 = 0.47$							

Table 27: Θ updated

E-step Since the labels of the data points were randomly initialized, they need to be updated given the latest estimation of Θ. We continue to update the labels of the data points. To facilitate the presentation, we invent a binary indicator variable, denoted as z_{nm}: $z_{nm} = 1$ indicates that the data point x_n was *assumed to be* sampled from the mth cluster; otherwise, $z_{nm} = 0$.

For example, if the first data point was sampled from the first cluster, the probability that $x_1 = 1.13$ is[149]

$$p(x_1 = 1.13|z_{11} = 1) = 0.22.$$

And if the first data point was sampled from the second cluster, the probability that $x_1 = 1.13$ is

$$p(x_1 = 1.13|z_{12} = 1) = 0.0003.$$

Repeat it for all the other data points, we have:

$$p(x_2 = 4.76|z_{21} = 1) = 0.05, p(x_2 = 4.76|z_{22} = 1) = 0.22;$$

$$p(x_3 = 0.87|z_{31} = 1) = 0.02, p(x_3 = 0.87|z_{32} = 1) = 0;$$

$$p(x_4 = 3.32|z_{41} = 1) = 0.15, p(x_4 = 3.32|z_{42} = 1) = 0.45;$$

$$p(x_5 = 4.29|z_{51} = 1) = 0.08, p(x_5 = 4.29|z_{52} = 1) = 0.45;$$

[149] In R, we could use the function `dnorm` to calculate it. For example, for $p(x_1 - 1.13|z_{11} = 1)$, we use `dnorm(1.13, mean = 1.75, sd = sqrt(2.83))` since $\mu_1 = 1.75, \sigma_1^2 = 2.83$.

$$p(x_6 = 1.03|z_{61} = 1) = 0.22, p(x_6 = 1.03|z_{62} = 1) = 0.0001;$$

$$p(x_7 = 0.98|z_{71} = 1) = 0.21, p(x_7 = 0.98|z_{72} = 1) = 0.0001.$$

Note that we need to calculate "the probability of *which cluster* a data point was sampled from"[150]. This is different from the probabilities we have calculated as shown above, which concerns "if a data point was sampled from a cluster, then the probability of the *specific value* the data point took on"[151].

Thus, we further calculate the conditional probabilities of $p(z_{i1}|x_i)$

$$p(z_{11} = 1|x_1 = 1.13) = \frac{0.22 \times 0.714}{0.22 \times 0.714 + 0.0003 \times 0.286} = 0.99; \text{ thus } x_1 \in C_1.$$

$$p(z_{21} = 1|x_2 = 4.76) = \frac{0.05 \times 0.714}{0.05 \times 0.714 + 0.22 \times 0.286} = 0.37; \text{ thus } x_2 \in C_2.$$

$$p(z_{31} = 1|x_3 = 0.87) = \frac{0.02 \times 0.714}{0.02 \times 0.714 + 0.00 \times 0.286} = 1; \text{ thus } x_3 \in C_1.$$

$$p(z_{41} = 1|x_4 = 3.32) = \frac{0.15 \times 0.714}{0.15 \times 0.714 + 0.45 \times 0.286} = 0.44; \text{ thus } x_4 \in C_2.$$

$$p(z_{51} = 1|x_5 = 4.29) = \frac{0.08 \times 0.714}{0.08 \times 0.714 + 0.45 \times 0.286} = 0.29; \text{ thus } x_5 \in C_2.$$

$$p(z_{61} = 1|x_6 = 1.03) = \frac{0.22 \times 0.714}{0.22 \times 0.714 + 0.0001 \times 0.286} = 0.99; \text{ thus } x_6 \in C_1.$$

$$p(z_{71} = 1|x_7 = 0.98) = \frac{0.21 \times 0.714}{0.21 \times 0.714 + 0.0001 \times 0.286} = 0.99; \text{ thus } x_7 \in C_1.$$

Table 27 can be updated to Table 28.

We can repeat this process and cycle through the two steps as shown in Figure 110, until the process converges, i.e., Θ remains the same (or its change is very small), or the labels of the data points remain the same. In this example, we actually only need one more iteration to reach convergence. This algorithm is a basic version of the so-called **EM algorithm**. Interested readers could find a complete derivation process in the **Remarks** section.

x_i	x_1	x_2	x_3	x_4	x_5	x_6	x_7
ID	1.13	4.76	0.87	3.32	4.29	1.03	0.98
Label	C1	C2	C1	C2	C2	C1	C1
$\Theta : \pi_1 = 0.57, \pi_2 = 0.43; \mu_1 = 1.00, \sigma_1^2 = 0.01; \mu_2 = 4.12, \sigma_2^2 = 0.54$							

Table 28: Cluster labels updated

Formal definition of the GMM

As a *data modeling* approach, the GMM implies a *data-generating mechanism*, that is summarized in below.

1. Suppose that there are M distributions mixed together.

2. In GMM, we assume that all the distributions are Gaussian distributions, i.e., the parameters of the mth distribution are $\{\mu_m, \Sigma_m\}$, and $m = 1, 2, \ldots, M$.[152]

   [152] μ_m is the mean vector; Σ_m is the covariance matrix.

3. For any data point x, without knowing its specific value, the prior probability that it comes from the mth distribution is denoted as π_m.[153] Note that $\sum_{m=1}^{M} \pi_m = 1$.

   [153] In other words, this is the percentage of the data points in the whole mix that come from the mth distribution.

The final distribution form of x is a mixed distribution with m components

$$x \sim \pi_1 N(\mu_1, \Sigma_1) + \pi_2 N(\mu_2, \Sigma_2) + \ldots + \pi_m N(\mu_m, \Sigma_m).$$

To learn the parameters from data, the EM algorithm is used. A basic walk-through of the EM algorithm has been given, i.e., see the example using Table 25.

R Lab

We simulate a dataset with 4 clusters as shown in Figure 111.

```
# Simulate a clustering structure
X <- c(rnorm(200, 0, 1), rnorm(200, 10,2),
       rnorm(200,20,1), rnorm(200,40, 2))
Y <- c(rnorm(800, 0, 1))
plot(X,Y, ylim = c(-5, 5), pch = 19, col = "gray25")
```

We use the R package `Mclust` to implement the GMM model using the EM algorithm.

```
# use GMM to identify the clusters
require(mclust)
XY.clust <- Mclust(data.frame(X,Y))
summary(XY.clust)
plot(XY.clust)
```

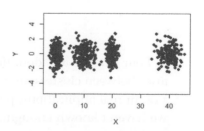

Figure 111: A mixture of 4 Gaussian distributions

We obtain the following result. Visualization of the identified clusters is shown in Figure 112. Note that we didn't specify the number of clusters in the analysis. `Mclust` used BIC and correctly identified the 4 clusters. For each cluster, the data points are about 200.

```
## --------------------------------------------------------
## Gaussian finite mixture model fitted by EM algorithm
## --------------------------------------------------------
##
## Mclust VVI (diagonal, varying volume and shape) model with
## 4 components:
##
##  log.likelihood  n df      BIC       ICL
##        -3666.07 800 19 -7459.147 -7459.539
##
## Clustering table:
##   1   2   3   4
## 199 201 200 200
```

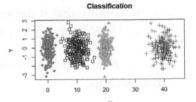

Figure 112: Clustering results of the simulated data

Now let's implement GMM on the AD data. Result is shown in Figure 113.

```
# install.packages("mclust")
require(mclust)
AD.Mclust <- Mclust(AD[,c(3,4,5,6,10,12,14,15)])
summary(AD.Mclust)
AD.Mclust$data = AD.Mclust$data[,c(1:4)]
# plot(AD.Mclust)
## --------------------------------------------------------
## Gaussian finite mixture model fitted by EM algorithm
## --------------------------------------------------------
##
## Mclust EEI (diagonal, equal volume and shape) model
## with 4 components:
##
##  log.likelihood  n df      BIC       ICL
##        -3235.874 517 43 -6740.414 -6899.077
##
## Clustering table:
##  1   2   3   4
## 43 253  92 129
```

Four clusters are identified as well. Figure 113 shows the boundaries between clusters are not as distinct as the boundaries in Figure 112. In real applications, particularly for those applications of which we haven't known enough, clustering is an exploration tool that could generate suggestive results but may not provide confirmatory conclusions.

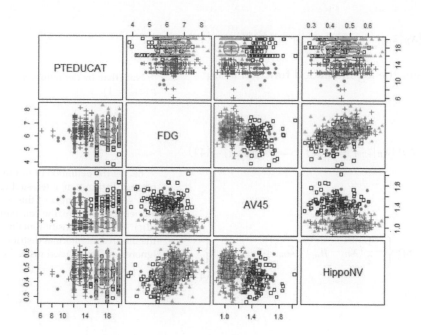

Figure 113: Clustering results of the AD dataset

Remarks

Derivation of the EM algorithm

The aforementioned two-step iterative algorithm (i.e., as outlined in Figure 110) illustrates how the **EM Algorithm** works. We have assumed that the two-step iterative algorithm would converge. Luckily, it had been proved that the EM Algorithm generally would converge[154].

The task of the EM algorithm is to learn the unknown parameters Θ from a given dataset. The Θ includes

[154] Wu, J., *On the Convergence Properties of the EM Algorithm*, The Annals of Statistics, Volume 11, Number 1, Pages 95-103, 1983.

1. The parameters of the M Gaussian distributions: $\{\mu_m, \Sigma_m, m = 1, 2, \ldots, M\}$.

2. The probability vector π that includes the elements $\{\pi_m, m = 1, 2, \ldots, M\}$.

Don't forget the binary indicator variable for each data point, denoted as z_{nm}: $z_{nm} = 1$ indicates that the data point x_n was sampled from the mth cluster[155].

The Likelihood Function   To learn these parameters from data, like in the logistic regression model, we derive a likelihood function to connect the data and parameters. For GMM, we cannot write $p(x_n|\Theta)$ directly. But it is possible to write $p(x_n, z_{nm}|\Theta)$ directly[156]

[155] The reason that z_{nm} is not included in Θ, as it could be seen later, after the presentation of the EM algorithm, is that z_{nm} provides a bridge to facilitate the learning of Θ. They are not essential parameters of the model, although they are useful to facilitate the estimation of the parameters of the model. Entities like z_{nm} are often called **latent variables** instead of *parameters*.

[156] That is what z_{nm} is needed for.

$$p\left(x_n, z_{nm}|\Theta\right) = \prod_{m=1}^{M} \left[p\left(x_n|z_{nm} = 1, \Theta\right) p\left(z_{nm} = 1\right)\right]^{z_{nm}}. \quad (40)$$

We apply *log* on Eq. 40 and get the log-likelihood function in Eq 41[157]

[157] Note that, by definition, $\pi_m = p\left(z_{nm} = 1\right)$.

$$\log p\left(x_n, z_{nm}|\Theta\right) = \sum_{m=1}^{M} \left[z_{nm} \log p\left(x_n|z_{nm} = 1, \Theta\right) + z_{nm} \log \pi_m\right]. \quad (41)$$

It is known that[158]

[158] I.e., by the definition of multivariate normal distribution; interested readers may see the **Appendix** of this book for a brief review. Here, the constant term $(2\pi)^{-p/2}$ in the density function of the multivariate normal distribution is ignored, so "$\propto$" is used instead of "$=$".

$$p\left(x_n|z_{nm} = 1, \Theta\right) \propto |\Sigma_m|^{-1/2} \exp\left\{-\frac{1}{2}\left(x_n - \mu_m\right)^T \Sigma_m^{-1} \left(x_n - \mu_m\right)\right\}. \quad (42)$$

Plug Eq. 42 into Eq. 41, we get

$$\log p\left(x_n, z_{nm}|\Theta\right) \propto$$

$$\sum_{m=1}^{M} \left[z_{nm}\left(-\frac{1}{2}\log|\Sigma_m| - \frac{1}{2}\left(x_n - \mu_m\right)^T \Sigma_m^{-1}\left(x_n - \mu_m\right) + z_{nm} \log \pi_m\right)\right]. \quad (43)$$

As there are N data points, the complete log-likelihood function is defined as

$$l(\Theta) = \log p(X, Z|\Theta) = \log \prod_{n=1}^{N} p\left(x_n, z_{nm}|\Theta\right). \quad (44)$$

With Eq. 43, Eq. 44 can be rewritten as

$$l(\Theta) \propto$$

$$\sum_{n=1}^{N}\sum_{m=1}^{M} \left[z_{nm}\left(-\frac{1}{2}\log|\Sigma_m| - \frac{1}{2}\left(x_n - \mu_m\right)^T \Sigma_m^{-1}\left(x_n - \mu_m\right) + z_{nm} \log \pi_m\right)\right]. \quad (45)$$

Optimization of $l(\Theta)$ Now we have an *explicit* form of $l(\Theta)$, based on which we use an optimization algorithm to search for the best estimate of Θ.

Recall that z_{nm} is unknown. Here comes the *initialization* again. Following the idea we have implemented in the data example shown in Table 25, we propose the following strategy:

1. Initialization. Either initialize $\{z_{nm}, n = 1, 2, \ldots, N; m = 1, 2, \ldots, M\}$ or Θ.

2. E-step. We can estimate z_{nm} if we have known Θ (i.e., given Θ), the best estimate of z_{nm} is the expectation of z_{nm} where the expectation is taken regarding the distribution $p\left(z_{nm}|x_n,\Theta\right)$ (i.e., denoted as $\langle Z_{nm}\rangle_{p(z_{nm}|x_n,\Theta)}$). By definition, we have

$$\langle z_{nm}\rangle_{p(z_{nm}|x_n,\Theta)} = 1 \cdot p\left(z_{nm}=1|x_n,\Theta\right) + 0 \cdot p\left(z_{nm}=0|x_n,\Theta\right).$$
(46)

It is known that

$$p\left(z_{nm}=1|x_n,\Theta\right) = \frac{p\left(x_n|z_{nm}=1,\Theta\right)\pi_m}{\sum_{k=1}^{M} p\left(x_n|z_{nk}=1,\Theta\right)\pi_k}.$$
(47)

Thus,

$$\langle z_{nm}\rangle_{p(z_{nm}|x_n,\Theta)} = \frac{p\left(x_n|z_{nm}=1,\Theta\right)\pi_m}{\sum_{k=1}^{M} p\left(x_n|z_{nk}=1,\Theta\right)\pi_k}.$$
(48)

3. M-step. Then, we derive the expectation of $l(\Theta)$ regarding the distribution $p\left(z_{nm}|x_n,\Theta\right)$

$$\langle l(\Theta)\rangle_{p(Z|X,\Theta)} = \sum_{n=1}^{N}\sum_{m=1}^{M}\Big[\langle z_{nm}\rangle_{p(z_{nm}=1|x_n,\Theta)}\log p\left(x_n|z_{nm}=1,\Theta\right) +$$
$$\langle z_{nm}\rangle_{p(z_{nm}=1|x_n,\Theta)}\log \pi_m\Big].$$
(49)

And we optimize Eq. 49 for Θ.

4. Repeat the E-step and M-step. With the updated Θ, we go back to the estimate of z_{nm} using Eq. 48, and then, feed the new estimate of z_{nm} into Eq. 49, and solve for Θ again. Repeat these iterations, until all the parameters in the iterations don't change significantly[159].

[159] Usually, we define a tolerance, e.g., the difference between two consecutive estimates of Θ is numerically bounded, such as 10^{-4}.

More about the M-step To estimate the parameters Θ, in the M-step we use the First Derivative Test again and take derivatives of $\langle l(\Theta)\rangle_{p(Z|X,\Theta)}$ (i.e., as shown in Eq. 49) regarding Θ and put the derivatives equal to zero.

For μ_m, we have

$$\frac{\partial\langle l(\Theta)\rangle_{p(Z|X,\Theta)}}{\partial\mu_m} = \sum_{n=1}^{N}\langle z_{nm}\rangle_{p(z_{nm}=1|x_n,\Theta)}\frac{\partial\log p\left(x_n|z_{nm}=1,\Theta\right)}{\partial\mu_m} = 0.$$
(50)

Based on Eq. 42, we can derive

$$\frac{\partial \log p\left(x_n | z_{nm} = 1, \Theta\right)}{\partial \mu_m} = -\frac{1}{2} \frac{\partial \left(x_n - \mu_m\right)^T \Sigma_m^{-1} \left(x_n - \mu_m\right)}{\partial \mu_m} = \left(x_n - \mu_m\right)^T \Sigma_m^{-1}.$$

(51)

Putting the result of Eq. 51 into Eq. 50, we can estimate μ_m by solving Eq. 50

$$\mu_m = \frac{\sum_{n=1}^{N} \langle z_{nm} \rangle_{p(z_{nm}=1|x_n,\Theta)} \, x_n}{\sum_{n=1}^{N} \langle z_{nm} \rangle_{p(z_{nm}=1|x_n,\Theta)}}.$$

(52)

Similarly, we take derivatives of $\langle l(\Theta) \rangle_{p(Z|X,\Theta)}$ regarding Σ_m and put the derivatives equal to zero

$$\frac{\partial \langle l(\Theta) \rangle_{p(Z|X,\Theta)}}{\partial \Sigma_m} = \sum_{n=1}^{N} \langle z_{nm} \rangle_{p(z_{nm}=1|x_n,\Theta)} \frac{\partial \log p\left(x_n | z_{nm} = 1, \Theta\right)}{\partial \Sigma_m} = 0.$$

(53)

Based on Eq. 42, we can derive

$$\frac{\partial \log p\left(x_n | z_{nm} = 1, \Theta\right)}{\partial \Sigma_m} =$$

$$\frac{1}{2} \frac{\partial \left\{ |\Sigma_m|^{-1/2} - \left(x_n - \mu_m\right)^T \Sigma_m^{-1} \left(x_n - \mu_m\right) \right\}}{\partial \Sigma_m} = \frac{1}{2} \left[\Sigma_m - \left(x_n - \mu_m\right) \left(x_n - \mu_m\right)^T \right].$$

(54)

Plug Eq. 54 into Eq. 53, we have

$$\sum_{n=1}^{N} \langle z_{nm} \rangle_{p(z_{nm}=1|X,\Theta)} \left[\Sigma_m - \left(x_n - \mu_m\right) \left(x_n - \mu_m\right)^T \right] = 0.$$

(55)

Solving Eq. 55, we estimate Σ_m as

$$\Sigma_m = \frac{\sum_{n=1}^{N} \langle z_{nm} \rangle_{p(z_{nm}=1|x_n,\Theta)} \left[\left(x_n - \mu_m\right) \left(x_n - \mu_m\right)^T \right]}{\sum_{n=1}^{N} \langle z_{nm} \rangle_{p(z_{nm}=1|x_n,\Theta)}}.$$

(56)

Lastly, to estimate π_m, recall that π_m is the percentage of the data points in the whole mix that come from the mth distribution, and $\pi_m = p\left(z_{nm} = 1\right)$, we can estimate π_m as

$$\pi_m = \frac{\sum_{n=1}^{N} \langle z_{nm} \rangle_{p(z_{nm}=1|x_n,\Theta)}}{N}.$$

(57)

Convergence of the EM Algorithm

Readers may have found that Eq. 45 gives us the form of $\log p(X, Z | \Theta)$, that is what is denoted as $l(\Theta)$. But, since Z is the latent variable and not part of the parameters, the objective function of the GMM model should be

$$\log p(X|\Theta) = \log \int p(X, Z|\Theta) dZ. \tag{58}$$

But this is not what has been done in the EM algorithm. Instead, the EM algorithm solves for Eq. 49, that is essentially

$$\langle \log p(X, Z|\Theta) \rangle_{p(Z|X,\Theta)} = \int \log p(X, Z; \Theta) p(Z|X, \Theta) dZ. \tag{59}$$

How does the solving of Eq. 59 help the solving of Eq. 58?

The power of the EM algorithm draws on **Jensen's inequality**. Let f be a convex function defined on an interval I. If $x_1, x_2, \ldots x_n \in I$ and $\gamma_1, \gamma_2, \ldots \gamma_n \geq 0$ with $\sum_{i=1}^{n} \gamma_i = 1$, then based on Jensen's inequality, it is known that $f\left(\sum_{i=1}^{n} \gamma_i x_i\right) \leq \sum_{i=1}^{n} \gamma_i f(x_i)$. Let's apply this result to analyze the EM algorithm.

First, notice that

$$\log p(X|\Theta) = \log \int p(X, Z|\Theta) dZ$$

$$= \log \int Q(Z) \frac{p(X, Z|\Theta)}{Q(Z)} dZ.$$

Here, $Q(Z)$ is any distribution of Z. In the EM algorithm

$$Q(Z) = p(Z|X, \Theta).$$

Using Jensen's inequality here, we have

$$\log \int Q(Z) \frac{p(X, Z|\Theta)}{Q(Z)} dZ$$

$$\geq \int Q(Z) \log \frac{p(X, Z|\Theta)}{Q(Z)} dZ.$$

Since

$$\int Q(Z) \log \frac{p(X, Z|\Theta)}{Q(Z)} dZ.$$

$$= \int Q(Z) \log p(X, Z|\Theta) dZ - \int Q(Z) Q(Z) dZ,$$

and $\int Q(Z) Q(Z) dZ$ is quadratic and thus non-negative, our final result is

$$\log p(X|\Theta) \geq \int Q(Z) \log p(X, Z|\Theta) dZ. \tag{60}$$

When we set $Q(Z) = p(Z|X, \Theta)$, Eq. 60 is rewritten as

$$\log p(X|\Theta) \geq \langle \log p(X, Z|\Theta)\rangle_{p(Z|X,\Theta)}. \tag{61}$$

Eq. 61 reveals that $\langle \log p(X, Z|\Theta)\rangle_{p(Z|X,\Theta)}$ is the **lower bound** of $\log p(X|\Theta)$. Thus, maximization of $\langle \log p(X, Z|\Theta)\rangle_{p(Z|X,\Theta)}$ can only increase the value of $\log p(X|\Theta)$. This is why solving Eq. 59 helps the solving of Eq. 58. This is the foundation of the effectiveness of the EM algorithm. The EM algorithm is often used to solve for problems that involve latent variables. Note that, $Q(Z)$ could be any distribution rather than $p(Z|X, \Theta)$, and Eq. 60 still holds. In applications where we could not explicitly derive $p(Z|X, \Theta)$, a surrogate distribution is used for $Q(Z)$. This variant of the EM algorithm is called the *variational inference*[160].

Clustering by random forest

Many clustering algorithms have been developed. The random forest model can be used for clustering as well. This is a byproduct utility of a random forest model. One advantage of using random forest for clustering is that it can cluster data points with mixed types of variables. To conduct clustering in random forests is to extract the distance information between data points that have been learned by the random forest model. There are multiple ways to do so. For example, one approach[161] that has been implemented in the R package `randomForests` is to generate a synthetic dataset with the same size as the original dataset, e.g., randomly generate the measurements of each variable using its empirical marginal distribution. The original dataset is taken as one class, while the synthetic dataset is taken as another class. Since the random forest model is used to classify the two classes, it will stress on the difference between the two datasets, which is, the variable dependency that is embedded in the original dataset but lost in the synthetic dataset because of the way the synthetic dataset is generated. Hence, each tree will be enriched with splitting variables that are dependent on other variables. After the random forest model is built, a distance between any pair of two data points can be calculated based on the frequency of this pair of data points existing in the same nodes of the random forest model. With this distance information, distance-based clustering algorithms such as the *hierarchical clustering* or *K-means clustering*[162] algorithms can be applied to detect the clusters.

In the following example, we generate a dataset with two clusters. The clusters produced from the random forest model are shown in Figure 114.

We then use the following R code to apply a random forest model

[160] A good starting point to know more about variational inference within a context of GMM, see: David, B., Kucukelbir, A. and McAuliffe, J., *Variational Inference: A Review for Statisticians*, Journal of the American Statistical Association, Volume 112, Number 518, Pages 859-877, 2017.

[161] Shi, T. and Horvath, S., *Unsupervised learning with random forest predictors.* Journal of Computational and Graphical Statistics, Volume 15, Issue 1, Pages 118-138, 2006.

[162] E.g., both could be implemented using the R package `cluster`.

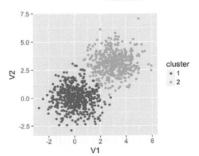

Figure 114: Clusters produced by the random forest model

on this dataset to find the clusters. It can be seen that the clusters are reasonably recovered by the random forest model.

```r
rm(list = ls(all = TRUE))
library(rpart)
library(dplyr)
library(ggplot2)
library(randomForest)
library(MASS)
library(cluster)
ndata <- 2000

sigma <- matrix(c(1, 0, 0, 1), 2, 2)
data1 <- mvrnorm(n = 500, rep(0, 2), sigma)
data2 <- mvrnorm(n = 500, rep(3, 2), sigma)
data <- rbind(data1, data2)
rf <- randomForest(data)
prox <- rf$proximity
clusters <- pam(prox, 2)
data <- as.data.frame(data)
data$cluster <- as.character(clusters$clustering)
ggplot(data, aes(x = V1, y = V2, color = cluster)) +
  geom_point() + labs(title = 'Data points')
```

Clustering-based prediction models

As we have mentioned, clustering is a flexible concept. And it could be used in a combination of methods. Figure 115 illustrates the basic idea of *clustering-based prediction models*. It applies a clustering algorithm first on the data and then builds a model for each cluster. As a data analytics strategy, we could combine different clustering algorithms and prediction models that are appropriate for an application context. There are also integrated algorithms that have articulated this strategy on the formulation level. For example, the *Treed Regression* method[163] is one example that proposed to build a tree to stratify the dataset first, and then, create regression models on the leaf nodes—here, each leaf node is a cluster. Similarly, the *logistic model trees*[164] also use a tree model to cluster data points into different leaf nodes and build different logistic regression model for each leaf node. Motivated by this line of thought, more models have been developed with different combination of tree models and prediction models (or other types of statistical models) on the leaf nodes[165] [166].

Exercises

1. In what follows is a summary of the clustering result on a dataset by using the R package `mclust`.

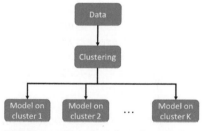

Figure 115: Clustering-based prediction models

[163] Alexander, W. and Grimshaw, S., *Treed regression.* Journal of Computational and Graphical Statistics, Volume 5, Issue 2, Pages 156-175, 1996.

[164] Landwehr, N., Hall, M. and Frank, E. *Logistic model trees.* Machine Learning, Volume 59, Issue 1, Pages 161–205, 2004.

[165] Gramacy, R. and Lee, H. *Bayesian treed Gaussian process models with an application to computer modeling.* Journal of American Statistical Association, Volume 103, Issue 483, Pages 1119-1130, 2008.

[166] Liu, H., Chen, X., Lafferty, J. and Wasserman, L. *Graph-valued regression.* In the Proceeding of Advances in Neural Information Processing Systems 23 (NIPS), 2010.

```
## ------------------------------------------------------
## Gaussian finite mixture model fitted by EM algorithm
## ------------------------------------------------------
##
## Mclust VVV (ellipsoidal, varying volume, shape,
## and orientation) model with 3
## components:
##
##  log-likelihood   n df      BIC       ICL
##      -2303.496 145 29 -4751.316 -4770.169
##
## Clustering table:
##  1  2  3
## 81 36 28
##
## Mixing probabilities:
##         1         2         3
## 0.5368974 0.2650129 0.1980897
##
## Means:
##              [,1]     [,2]      [,3]
## glucose  90.96239 104.5335  229.42136
## insulin 357.79083 494.8259 1098.25990
## sspg    163.74858 309.5583   81.60001
##
## Variances:
## [,,1]
##           glucose    insulin       sspg
## glucose  57.18044   75.83206   14.73199
## insulin  75.83206 2101.76553  322.82294
## sspg     14.73199  322.82294 2416.99074
## [,,2]
##           glucose   insulin       sspg
## glucose  185.0290  1282.340  -509.7313
## insulin 1282.3398 14039.283 -2559.0251
## sspg    -509.7313 -2559.025 23835.7278
## [,,3]
##           glucose   insulin       sspg
## glucose  5529.250  20389.09  -2486.208
## insulin 20389.088  83132.48 -10393.004
## sspg    -2486.208 -10393.00   2217.533
```

2. (a) How many samples in total are in this dataset? How many variables? (b) How many clusters are found? What are the sizes of the clusters? (c) What is the fitted GMM model? Please write its mathematical form.

3. Consider the dataset in Table 29 that has 9 data points. Let's use it to estimate a GMM model with 3 clusters. The initial values are shown in Table 29

ID	1.53	0.57	2.56	1.22	4.13	6.03	0.98	5.21	−0.37
Label	C1	C3	C1	C2	C2	C2	C1	C2	C3

Table 29: Initial values for a GMM model with 3 clusters

(a) Write the Gaussian mixture model (GMM) that you want to estimate. (b) Estimate the parameters of your GMM model. (c) Update the labels with your estimated parameters. (d) Estimate the parameters again.

4. Follow up on the dataset in Q3. Use the R pipeline for clustering on this data. Compare the result from R and the result from your manual calculation.

5. Consider the dataset in Table 30 that has 10 data points. Let's use it to estimate a GMM model with 3 clusters. The initial values are shown in Table 30

ID	2.22	6.33	3.15	−0.89	3.21	1.10	1.58	0.03	8.05	0.26
Label	C1	C3	C1	C2	C2	C2	C1	C2	C3	C2

Table 30: Initial values for a GMM model with 3 clusters

(a) Write the Gaussian mixture model (GMM) that you want to estimate. (b) Estimate the parameters of your GMM model. (c) Update the labels with your estimated parameters. (d) Estimate the parameters again.

6. Design a simulation experiment to test the effectiveness of the R package `mclust`. For instance, simulate a three-cluster structure in your dataset by this GMM model

$$x \sim \pi_1 N\left(\mu_1, \Sigma_1\right) + \pi_2 N\left(\mu_2, \Sigma_2\right) + \pi_3 N\left(\mu_3, \Sigma_3\right),$$

where $\pi_1 = 0.5$, $\pi_2 = 0.25$, and $\pi_3 - 0.25$, and

$$\mu_1 = \begin{bmatrix} 5 \\ 3 \\ 3 \end{bmatrix}, \mu_2 = \begin{bmatrix} 10 \\ 5 \\ 1 \end{bmatrix}, \mu_3 = \begin{bmatrix} -5 \\ 10 \\ -2 \end{bmatrix};$$

$$\Sigma_1 = \begin{bmatrix} 1 & 0 & 0 \\ 0 & 1 & 0 \\ 0 & 0 & 1 \end{bmatrix}, \Sigma_2 = \begin{bmatrix} 1 & 0 & 0 \\ 0 & 1 & 0 \\ 0 & 0 & 1 \end{bmatrix}, \Sigma_3 = \begin{bmatrix} 1 & 0 & 0 \\ 0 & 1 & 0 \\ 0 & 0 & 1 \end{bmatrix}.$$

Then, use the `mclust` package on this dataset and see if the true clustering structure could be recovered.

7. Follow up on the simulation experiment in Q6. Let's consider a GMM model with larger variance

$$x \sim \pi_1 N\left(\mu_1, \Sigma_1\right) + \pi_2 N\left(\mu_2, \Sigma_2\right) + \pi_3 N\left(\mu_3, \Sigma_3\right),$$

where $\pi_1 = 0.5$, $\pi_2 = 0.25$, and $\pi_3 = 0.25$, and

$$\mu_1 = \begin{bmatrix} 5 \\ 3 \\ 3 \end{bmatrix}, \; \mu_2 = \begin{bmatrix} 10 \\ 5 \\ 1 \end{bmatrix}, \; \mu_3 = \begin{bmatrix} -5 \\ 10 \\ -2 \end{bmatrix};$$

$$\Sigma_1 = \begin{bmatrix} 3 & 0 & 0 \\ 0 & 3 & 0 \\ 0 & 0 & 3 \end{bmatrix}, \; \Sigma_2 = \begin{bmatrix} 3 & 0 & 0 \\ 0 & 3 & 0 \\ 0 & 0 & 3 \end{bmatrix}, \; \Sigma_3 = \begin{bmatrix} 3 & 0 & 0 \\ 0 & 3 & 0 \\ 0 & 0 & 3 \end{bmatrix}.$$

Then, use the R package `mclust` on this dataset and see if the true clustering structure could be recovered.

8. Design a simulation experiment to test the effectiveness of the diagnostic tools in the `ggfortify` R package. For instance, use the same simulation procedure that has been used in Q9 of **Chapter 2** to design a linear regression model with two variables, simulate 100 samples from this model, fit the model, and draw the diagnostic figures.

9. Follow up on the simulation experiment in Q8. Add a few outliers into your dataset and see if the diagnostic tools in the `ggfortify` R package can detect them.

Chapter 7: Learning (II)
SVM & Ensemble Learning

Overview

Chapter 7 revisits *learning* from a perspective that is different from
Chapter 5. In **Chapter 5** we have introduced the concept of overfitting and the use of cross-validation as a safeguard mechanism to
help us build models that don't overfit the data. It focused on fair
evaluation of the performances of a *specific model*. **Chapter 7**, taking on a process-oriented view of the issue of overfitting, focuses on
performances of a *learning algorithm*[167]. This chapter introduces two
methods that aim to build a safeguard mechanism into the learning
algorithms themselves. The two methods are the **Support Vector Machine (SVM)** and **Ensemble Learning**[168]. While all models could
overfit a dataset, these methods aim to reduce risk of overfitting
based on their unique modeling principles.

    In short, **Chapter 5** introduced evaluative methods that concern *if a
model has learned from the data*. It is about quality assessment. **Chapter
7** introduces learning methods that concern *how to learn better from the
data*. It is about quality improvement.

[167] Algorithms are computational
procedures that learn models from data.
They are processes.

[168] The random forest model is a typical
example of ensemble learning

Support vector machine

Rationale and formulation

A learning algorithm has an *objective function* and sometimes a set
of *constraints*. The objective function corresponds to a quality of the
learned model that could help it succeed on the unseen testing data.
Eqs. 16, 30, and 45, are examples of objective functions. They are
developed based on the *likelihood principle*. Besides the likelihood
principle, researchers have been studying what else quality a model
should have and what objective function we should optimize to enhance this quality of the model. The constraints, on the other hand,
guard the bottom line: the learned model needs to at least perform

well on the training data so it is possible to perform well on future unseen data[169].

Figure 116 shows an example of a binary classification problem. The constraints here are obvious: the models should correctly classify the data points. And the 3 models all perform well, while we hesitate to say that the 3 models are equally good. Common sense tells us that Model 3 is the least favorable. Unlike the other two, Model 3 is close to a few data points. This makes Model 3 bear a risk of misclassification on future unseen data: the locations of the existing data points provide a suggestion about where future unseen data may locate; but this is a suggestion, not a hard boundary.

In other words, the line of Model 3 is too close to the data points and therefore lacks a safe **margin**. The concept of margin is shown in Figure 117. To reduce risk, we should have the margin as large as possible. The other two models have larger margins, and Model 2 is the best because it has the largest margin.

In summary, while all the models shown in Figure 116 meet the *constraints* (i.e., perform well on the training data points), this is just the bottom line for a model to be good, and they are ranked differently based on an *objective function* that maximizes the margin of the model. This is the **maximum margin** principle invented in SVM.

Theory and method

Derivation of the SVM formulation Consider a binary classification problem as shown in Figure 117. At this moment, we consider situations that all data points could be correctly classified by a line, which is clearly the case in Figure 116. This is called **the linearly separable case**. Denote the data points as $\{(x_n, y_n), n = 1, 2, \ldots, N\}$. Here, the outcome variable y is denoted as $y_n \in \{1, -1\}$, i.e., $y = 1$ denotes the circle points; $y = -1$ denotes the square points.

The mathematical model to represent a line is $w^T x + b = 0$. Based on this form, we can segment the space into 5 regions, as shown in Figure 118. And by looking at the value of $w^T x + b$, we know which region the data point x falls into. In other words, Figure 118 tells us a *classification rule*

$$\text{If } w^T x + b > 0, \text{ then } y = 1;$$
$$\text{Otherwise, } y = -1. \tag{62}$$

Note that

$$\text{For data points on the margin: } \left| w^T x + b \right| = 1;$$
$$\text{For data points beyond the margin: } \left| w^T x + b \right| > 1. \tag{63}$$

[169] The testing data, while unseen, is assumed to be statistically the same as the training data. This is a basic assumption in machine learning.

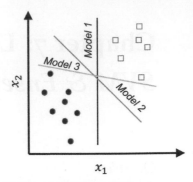

Figure 116: Which model (i.e., here, which line) should we use as our classification model to separate the two classes of data points?

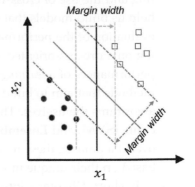

Figure 117: The model that has a larger margin is better—the basic idea of SVM

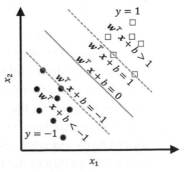

Figure 118: The 5 regions

These two equations in Eq. 63 provide the *constraints* for the SVM formulation, i.e., the bottom line for a model to be a good model. The two equations can be succinctly rewritten as one

$$y\left(w^T x + b\right) \geq 1.$$

Thus, a draft version of the SVM formulation is

Objective function: Maximize Margin,

Subject to: $y_n\left(w^T x_n + b\right) \geq 1$ for $n = 1, 2, \ldots, N$. $\quad$ (64)

The *objective function* is to maximize the *margin* of the model. Note that a model is characterized by its parameters w and b. And the goal of Eq. 64 is to find the model—and therefore, the parameters—that maximizes the margin. In order to carry out this idea, we need the margin to be a concrete mathematical entity that *could be* characterized by the parameters w and b[170].

We refer readers to the **Remarks** section to see details of how the margin is derived as a function of w. Figure 119 shows the result: the margin of the model is $\frac{2}{\|w\|}$. Here, $\|w\|^2 = w^T w$. And note that to *maximize the margin of a model* is equivalent to *minimize* $\|w\|$. This gives us the objective function of the SVM model[171]

$$\text{Maximize Margin} = \min_w \frac{1}{2}\|w\|^2. \quad (65)$$

Thus, the final SVM formulation is

$$\min_w \frac{1}{2}\|w\|^2,$$
$$\text{Subject to: } y_n\left(w^T x_n + b\right) \geq 1 \text{ for } n = 1, 2, \ldots, N. \quad (66)$$

Optimization solution   Eq. 66 is called the **primal formulation** of SVM. To solve it, it is often converted into its dual form, the **dual formulation** of SVM. This could be done by the method of **Lagrange multiplier** that introduces a dummy variable, α_n, for each constraint, i.e., $y_n\left(w^T x_n + b\right) \geq 1$, such that we could move the constraints into the objective function. By definition, $\alpha_n \geq 0$.

$$L(w, b, \alpha) = \frac{1}{2}\|w\|^2 - \sum_{n=1}^{N} \alpha_n \left[y_n\left(w^T x_n + b\right) - 1\right].$$

This could be rewritten as

$$L(w, b, \alpha) = \underbrace{\frac{1}{2} w^T w}_{(1)} - \underbrace{\sum_{n=1}^{N} \alpha_n y_n w^T x_n}_{(2)} - \underbrace{b \sum_{n=1}^{N} \alpha_n y_n}_{(3)} + \underbrace{\sum_{n=1}^{N} \alpha_n}_{(4)}. \quad (67)$$

[170] Not all good ideas could be readily materialized in concrete mathematical forms. There is no guaranteed mathematical reality and if there is one it is always hard-earned.

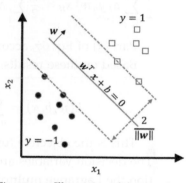

Figure 119: Illustration of the margin as a function of w

[171] Note that here we use $\|w\|^2$ instead of $\|w\|$. This formulation is easier to solve.

Then we use the First Derivative Test again: differentiating $L(w, b, \alpha)$ with respect to w and b, and setting them to 0 yields the following solutions

$$w = \sum_{n=1}^{N} \alpha_n y_n x_n; \tag{68}$$

$$\sum_{n=1}^{N} \alpha_n y_n = 0. \tag{69}$$

Using the conclusion in Eq. 68, part (1) of Eq. 67 could be rewritten as

$$\frac{1}{2} w^T w = \frac{1}{2} w^T \sum_{n=1}^{N} \alpha_n y_n x_n = \frac{1}{2} \sum_{n=1}^{N} \alpha_n y_n w^T x_n.$$

It has the same form as part (2) of Eq. 67. The two could be merged together into $-\frac{1}{2} \sum_{n=1}^{N} \alpha_n y_n w^T x_n$. Note that[172]

[172] I.e., use the conclusion in Eq. 68 again.

$$\frac{1}{2} \sum_{n=1}^{N} \alpha_n y_n w^T x_n = \frac{1}{2} \sum_{n=1}^{N} \alpha_n y_n \left(\sum_{n=1}^{N} \alpha_n y_n x_n \right)^T x_n = \frac{1}{2} \sum_{n=1}^{N} \sum_{m=1}^{N} \alpha_n \alpha_m y_n y_m x_n^T x_m.$$

Part (3) of Eq. 67, according to the conclusion in Eq. 69, is 0. Based on these results, we can rewrite $L(w, b, \alpha)$ as

$$L(w, b, \alpha) = \sum_{n=1}^{N} \alpha_n - \frac{1}{2} \sum_{n=1}^{N} \sum_{m=1}^{N} \alpha_n \alpha_m y_n y_m x_n^T x_m.$$

This is the objective function of the dual formulation of Eq. 66. The decision variables are the Lagrange multipliers, the α. By definition the Lagrange multipliers should be non-negative, and we have the constraint of the Lagrange multipliers described in Eq. 69. All together, the **dual formulation** of the SVM model is

$$\max_{\alpha} \sum_{n=1}^{N} \alpha_n - \frac{1}{2} \sum_{n=1}^{N} \sum_{m=1}^{N} \alpha_n \alpha_m y_n y_m x_n^T x_m, \tag{70}$$

Subject to: $\alpha_n \geq 0$ for $n = 1, 2, \ldots, N$, and $\sum_{n=1}^{N} \alpha_n y_n = 0$.

This is a *quadratic programming* problem that can be solved using many existing well established algorithms.

Support vectors   The data points that lay on the margins, as shown in Figure 120, are called **support vectors**. These geometrically unique data points are also found to be numerically interesting: in the solution of the dual formulation of SVM as shown in Eq. 70, the α_ns that correspond to the support vectors are those that are nonzero. In other words, the data points that are not support vectors will have their α_ns to be zero in the solution of Eq. 70.[173]

If we revisit Eq. 68, we can see that only the nonzero α_n contribute to the estimation of w. Indeed, Figure 120 shows that support vectors are sufficient to geometrically define the margins. And if we know the margins, the decision boundary is determined, i.e., as the central line in the middle of the two margins.

The support vectors hold crucial implications for the learned model. Theoretical evidences showed that the number of support vectors is a metric that can indicate the "healthiness" of the model, i.e., the smaller the total number of support vectors, the better the model. It also reveals that the main statistical information of a given dataset the SVM model uses is the support vectors. The number of support vectors is usually much smaller than the number of data points N. Some works have been inspired to accelerate the SVM model training by discarding the data points that are probably not support vectors[174]. To understand why the nonzero α_n correspond to the support vectors, interested readers can find the derivation in the **Remarks** section.

Summary After solving Eq. 70, we obtain the solutions of α. With that, we estimate the parameter w based on Eq. 68. To estimate the parameter b, we use any *support vector*, i.e., say, (x_n, y_n), and estimate b by

$$\text{If } y_n = 1, b = 1 - w^T x_n;$$

$$\text{If } y_n = -1, b = -1 - w^T x_n. \tag{71}$$

Extension to nonseparable cases We have assumed that the two classes are separable. Since this is impossible in some applications, we revise the SVM formulation—specifically, to revise the constraints of the SVM formulation—by allowing some data points to be within the margins or even on the wrong side of the decision boundary.

Note that the original constraint structure in Eq. 66 is derived based on the linearly separable case shown in Figure 116. For the nonseparable case, Figure 121 shows three scenarios: the *Type A* data points fall within the margins but still on the right side of their class, the *Type B* data points fall on the wrong side of their class, and the *Type C* data points fall on the right side of their class and also beyond or on the margin.

The *Type A* data points and the *Type B* data points are both *compromised*, and we introduce a **slack variable** to describe the *degree* of

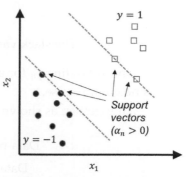

Figure 120: Support vectors are the data points that lay on the margins. In other words, the support vectors define the margins.

[174] If we can screen the data points before we solve Eq. 70 by discarding some data points that are not support vectors, the size of the optimization problem in Eq. 70 could be reduced.

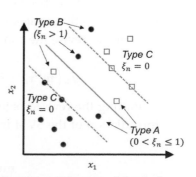

Figure 121: Behaviors of the slack variables

compromise for both types of data points.

For instance, consider the circle points that belong to the class $(y_n = 1)$, we have[175]

[175] Readers may revisit Figure 118 to understand Eq. 72.

$$\text{Data points (Type A): } w^T x_n + b \in (0,1);$$
$$\text{Data points (Type B): } w^T x_n + b < 0. \tag{72}$$

Then we define a slack variable ξ_n for any data point n of Types A or B

$$\text{The slack variable } \xi_n : \xi_n = 1 - \left(w^T x_n + b\right).$$

And we define ξ_n for any data point of Type C to be 0 since there is no compromise.

All together, as shown in Figure 121, we have

$$\text{Data points (Type A): } \xi_n \in (0,1];$$
$$\text{Data points (Type B): } \xi_n > 1; \tag{73}$$
$$\text{Data points (Type C): } \xi_n = 0.$$

Similarly, for the square points that belong to the class $(y = -1)$, we define a slack variable ξ_n for each data point n

$$\text{The slack variable } \xi_n : \xi_n = 1 + \left(w^T x_n + b\right).$$

The same result in Eq. 73 could be derived.

As the slack variable ξ_n describes the *degree* of compromise for the data point x_n, an optimal SVM model should also minimize the total amount of compromise. Based on this additional learning principle, we revise the objective function in Eq. 66 and get

$$\underbrace{\min_{w} \frac{1}{2}\|w\|^2}_{\text{Maximize Margin}} + \underbrace{C \sum_{n=1}^{N} \xi_n.}_{\text{Minimize Slacks}} \tag{74}$$

Here, C is a user-specified parameter to control the balance between the two objectives: *maximum margin* and *minimum sum of slacks*.

[176] I.e., use the results in Figure 118 and Figure 121.

Then we revise the constraints[176] to be

$$y_n \left(w^T x_n + b\right) \geq 1 - \xi_n, \text{ for } n = 1, 2, \ldots, N.$$

Putting the revised objective function and constraints together, the formulation of the SVM model for nonseparable case becomes

$$\min_{w} \frac{1}{2}\|w\|^2 + C \sum_{n=1}^{N} \xi_n,$$
$$\text{Subject to: } y_n \left(w^T x_n + b\right) \geq 1 - \xi_n, \tag{75}$$
$$\xi_n \geq 0, \text{ for } n = 1, 2, \ldots, N.$$

A dual form that is similar to Eq. 70 could be derived, which is skipped here[177].

Extension to nonlinear SVM   Sometimes, the decision boundary could not be characterized as linear models, i.e., see Figure 122 (a).

[177] Interested readers could read this book for a comprehensive and deep understanding of SVM: Scholkopf, B. and Smola, A.J., *Learning with Kernels: Support Vector Machines, Regularization, Optimization, and Beyond*. MIT Press, 2001.

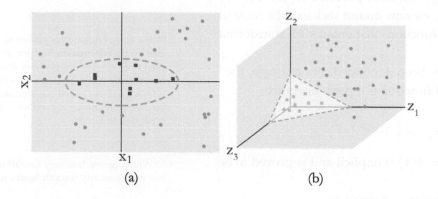

(a)   (b)

Figure 122: (a) A nonseparable dataset; (b) with the right transformation, (a) becomes linearly separable

A common strategy to create a nonlinear model is to conduct *transformation* of the original variables. For Figure 122 (a), we conduct a transformation from the original two-dimensional coordinate system x to a new coordinate system z that is three-dimensional

$$z_1 = x_1^2, z_2 = \sqrt{2}x_1x_2, z_3 = x_2^2. \tag{76}$$

In the new coordinate system, as shown in Figure 122 (b), the data points of the two classes become linearly separable.

The transformation employed in Eq. 76 is *explicit*, which may not be suitable for applications where we don't know what is a good transformation[178]. Thus, transformation that could be automatically identified by the learning algorithm is needed, even if the transformation is *implicit*. A remarkable thing about SVM is that its formulation allows automatic transformation.

Let's revisit the dual formulation of SVM for the linearly separable case, as shown in Eq. 70. Assume that the transformation has been performed and now we build the SVM model based on the transformed features, z. The dual formulation of SVM on the transformed variables is

[178] Try a ten-dimensional x and see how troublesome it is to define an explicit transformation to enable linear separability of the classes.

$$\max_{\alpha} \sum_{n=1}^{N} \alpha_n - \frac{1}{2} \sum_{n=1}^{N} \sum_{m=1}^{N} \alpha_n \alpha_m y_n y_m z_n^T z_m,$$

$$\text{Subject to: } \alpha_n \geq 0 \text{ for } n = 1, 2, \ldots, N, \tag{77}$$

$$\sum_{n=1}^{N} \alpha_n y_n = 0.$$

It can be seen that, the dual formulation of SVM doesn't directly concern z_n. Rather, only the inner product of $z_n^T z_m$ is needed. As z is essentially a function of x, i.e., denote it as $z = \phi(x)$, $z_n^T z_m$ is essentially a function of x_n and x_m. We can write it up as $z_n^T z_m = K(x_n, x_m)$. This is called the **kernel function**.

A kernel function is a function that entails a transformation $z = \phi(x)$ such that $K(x_n, x_m)$ is an inner product: $K(x_n, x_m) = \phi(x_n)^T \phi(x_m)$. In other words, we now do not seek explicit form of $\phi(x_n)$; rather, we seek kernel functions that entail such transformations[179].

Many kernel functions have been developed. For example, the **Gaussian radial basis kernel function** is a popular choice

$$K(x_n, x_m) = e^{-\gamma \|x_n - x_m\|^2},$$

where the transformation $z = \phi(x)$ is implicit and is proved to be infinitely long[180].

The polynomial kernel function is defined as

$$K(x_n, x_m) = \left(x_n^T x_m + 1\right)^q.$$

The linear kernel function[181] is defined as

$$K(x_n, x_m) = x_n^T x_m.$$

With a given kernel function, the dual formulation of SVM is

$$\max_{\alpha} \sum_{n=1}^{N} \alpha_n - \frac{1}{2} \sum_{n=1}^{N} \sum_{m=1}^{N} \alpha_n \alpha_m y_n y_m K(x_n, x_m),$$

Subject to: $\alpha_n \geq 0$ for $n = 1, 2, \ldots, N,$ \hfill (78)

$$\sum_{n=1}^{N} \alpha_n y_n = 0.$$

After solving Eq. 78, *in theory* we could obtain the estimation of the parameter w based on Eq. 79

$$w = \sum_{n=1}^{N} \alpha_n y_n \phi(x_n). \hfill (79)$$

However, for kernel functions that we don't know the explicit transformation function $\phi(x)$, it is no longer possible to write the parameter w in the same way as in linear SVM models. This won't prevent us from using the learned SVM model for prediction. For a data point, denoted as x_*, we can use the learned SVM model to predict on it[182]

[179] If a kernel function is proven to entail a transformation function $\phi(x)$— even it is only proven *in theory* and never really made explicit in practice— it is as good as explicit transformation, because only the inner product of $z_n^T z_m$ is needed in Eq. 77.

[180] Which means it is very flexible and can represent any smooth function.

[181] For linear kernel function, the transformation is trivial, i.e., $\phi(x) = x$.

[182] I.e., combine Eq. 79 and Eq. 62 we could derive Eq. 80.

If $\sum_{n=1}^{N} \alpha_n y_n K\left(x_n, x_*\right) + b > 0$, then $y_* = 1$;

$$\text{(80)}$$

Otherwise, $y_* = -1$.

Again, the specific form of $\phi(x)$ is not needed since only the kernel function is used.

A small-data example   Consider a dataset with 4 data points

$$
\begin{aligned}
x_1 &= (-1, -1)^T, y_1 = -1;\\
x_2 &= (-1, +1)^T, y_2 = +1;\\
x_3 &= (+1, -1)^T, y_3 = +1;\\
x_4 &= (+1, +1)^T, y_4 = -1.
\end{aligned}
$$

The dataset is visualized in Figure 123. The R code to draw Figure 123 is shown below.

```
# For the toy problem
x = matrix(c(-1,-1,1,1,-1,1,-1,1), nrow = 4, ncol = 2)
y = c(-1,1,1,-1)
linear.train <- data.frame(x,y)

# Visualize the distribution of data points of two classes
require( 'ggplot2' )
p <- qplot( data=linear.train, X1, X2,
            colour=factor(y),xlim = c(-1.5,1.5),
            ylim = c(-1.5,1.5))
p <- p + labs(title = "Scatterplot of data points of two classes")
print(p)
```

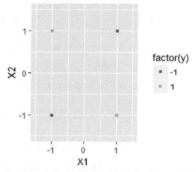

Figure 123: A linearly inseparable dataset

It is a *nonlinear* case. We use a nonlinear kernel function to build the SVM model.

Consider the polynomial kernel function with `df=2`

$$
K\left(x_n, x_m\right) = \left(x_n^T x_m + 1\right)^2,
\tag{81}
$$

which corresponds to the transformation

$$
\phi\left(x_n\right) = \left[1, \sqrt{2}x_{n,1}, \sqrt{2}x_{n,2}, \sqrt{2}x_{n,1}x_{n,2}, x_{n,1}^2, x_{n,2}^2\right]^T.
\tag{82}
$$

Based on Eq. 70, a specific formulation of the SVM model of this dataset is

$$
\max_{\alpha} \sum_{n=1}^{4} \alpha_n - \frac{1}{2} \sum_{n=1}^{4} \sum_{m=1}^{4} \alpha_n \alpha_m y_n y_m K\left(x_n, x_m\right),
$$

$$\text{Subject to: } \alpha_n \geq 0 \text{ for } n = 1, 2, \ldots, 4,
\tag{83}$$

$$\text{and } \sum_{n=1}^{4} \alpha_n y_n = 0.$$

We calculate the kernel matrix as[183]

$$K = \begin{bmatrix} 9 & 1 & 1 & 1 \\ 1 & 9 & 1 & 1 \\ 1 & 1 & 9 & 1 \\ 1 & 1 & 1 & 9 \end{bmatrix}.$$

We solve the quadratic programming problem[184] in Eq. 83 and get

$$\alpha_1 = \alpha_2 = \alpha_3 = \alpha_4 = 0.125. \tag{84}$$

In this particular case, since we can write up the transformation explicitly[185], we can write up w explicitly as well[186]

$$w = \sum_{n=1}^{4} \alpha_n y_n \phi(x_n) = [0, 0, 0, 1/\sqrt{2}, 0, 0]^T.$$

For any given data point x_*, the explicit decision function is

$$f(x_*) = w^T \phi(x_*) = x_{*,1} x_{*,2}.$$

This is the decision boundary for a typical **XOR** problem[187].

We then use R to build an SVM model on this dataset[188]. The R code is shown in below.

```
# Train a nonlinear SVM model
# polynomial kernel function with 'df=2'
x <- cbind(1, poly(x, degree = 2, raw = TRUE))
coefs = c(1,sqrt(2),1,sqrt(2),sqrt(2),1)
x <- x * t(matrix(rep(coefs,4),nrow=6,ncol=4))
linear.train <- data.frame(x,y)
require( 'kernlab' )
linear.svm <- ksvm(y ~ ., data=linear.train,
                type='C-svc', kernel='vanilladot', C=10, scale=c())
```

The function `alpha()` returns the values of α_n for $n = 1, 2, \ldots, 4$. Our results as shown in Eq. 84 are consistent with the results obtained by using R.[189]

```
alpha(linear.svm) #scaled alpha vector
## [[1]]
## [1] 0.125 0.125 0.125 0.125
```

[183] E.g., using Eq. 81, $K(x_1, x_2) = (x_1^T x_2 + 1)^2 = 3^2 = 9$. Readers can try other instances.

[184] I.e., use the R package `quadprog`.

[185] I.e., as shown in Eq. 82

[186] It should be written as $\hat{w}$, since it is an estimator of w. Here for simplicity we skip this.

[187] Also known as *exclusive or* or *exclusive disjunction*, the XOR problem is a logical operation that outputs *true* only when inputs differ (e.g., one is *true*, the other is *false*).

[188] We use the R package `kernlab` —more details are shown in the section **R Lab**.

[189] If your answer is different, check if the `alpha()` function in the `kernlab` () package scales the vector α, i.e., to make the sum as 1.

R Lab

The 7-Step R Pipeline   **Step 1** and **Step 2** get data into R and make appropriate preprocessing.

```r
# Step 1 -> Read data into R workstation

library(RCurl)
url <- paste0("https://raw.githubusercontent.com",
              "/analyticsbook/book/main/data/AD.csv")
data <- read.csv(text=getURL(url))

# Step 2 -> Data preprocessing
# Create X matrix (predictors) and Y vector (outcome variable)
X <- data[,2:16]
Y <- data$DX_bl

Y <- paste0("c", Y)
Y <- as.factor(Y)

data <- data.frame(X,Y)
names(data)[16] = c("DX_bl")

# Create a training data (half the original data size)
train.ix <- sample(nrow(data),floor( nrow(data)/2) )
data.train <- data[train.ix,]
# Create a testing data (half the original data size)
data.test <- data[-train.ix,]
```

Step 3 puts together a list of candidate models.

```r
# Step 3 -> gather a list of candidate models
# SVM: often to compare models with different kernels,
# different values of C, different set of variables

# Use different set of variables

model1 <- as.formula(DX_bl ~ .)
model2 <- as.formula(DX_bl ~ AGE + PTEDUCAT + FDG
                     + AV45 + HippoNV + rs3865444)
model3 <- as.formula(DX_bl ~ AGE + PTEDUCAT)
model4 <- as.formula(DX_bl ~ FDG + AV45 + HippoNV)
```

Step 4 uses 10-fold cross-validation to evaluate the performance of the candidate models. Below we show how it works for one model. For other models, the same script could be used with a slight modification.

```r
# Step 4 -> Use 10-fold cross-validation to evaluate the models

n_folds = 10
```

```
# number of fold
N <- dim(data.train)[1]
folds_i <- sample(rep(1:n_folds, length.out = N))

# evaluate the first model
cv_err <- NULL
# cv_err makes records of the prediction error for each fold
for (k in 1:n_folds) {
  test_i <- which(folds_i == k)
  # In each iteration, use one fold of data as the testing data
  data.test.cv <- data.train[test_i, ]
  # The remaining 9 folds' data form our training data
  data.train.cv <- data.train[-test_i, ]
  require( 'kernlab' )
  linear.svm <- ksvm(model1, data=data.train.cv,
                     type='C-svc', kernel='vanilladot', C=10)
  # Fit the linear SVM model with the training data
  y_hat <- predict(linear.svm, data.test.cv)
  # Predict on the testing data using the trained model
  true_y <- data.test.cv$DX_bl
  # get the the error rate
  cv_err[k] <-length(which(y_hat != true_y))/length(y_hat)
}
mean(cv_err)

# evaluate the second model ...
# evaluate the third model ...
# ...
```

Results are shown below.

```
## [1] 0.1781538
## [1] 0.1278462
## [1] 0.4069231
## [1] 0.1316923
```

The second model is the best.

Step 5 uses the training data to fit a final model, through the `ksvm()` function in the package `kernlab`.

```
# Step 5 -> After model selection,
# use ksvm() function to build your final model
linear.svm <- ksvm(model2, data=data.train,
        type='C-svc', kernel='vanilladot', C=10)
```

Step 6 uses the fitted final model for prediction on the testing data.

```
# Step 6 -> Predict using your SVM model
y_hat <- predict(linear.svm, data.test)
```

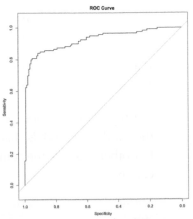

Figure 124: The ROC curve of the final SVM model

Step 7 evaluates the performance of the model.

```
# Step 7 -> Evaluate the prediction performance of the SVM model

# (1) The confusion matrix

library(caret)
confusionMatrix(y_hat, data.test$DX_bl)

# (2) ROC curve
library(pROC)
y_hat <- predict(linear.svm, data.test, type = 'decision')
plot(roc(data.test$DX_bl, y_hat),
     col="blue", main="ROC Curve")
```

Results are shown below. And the ROC curve is shown in Figure 124.

```
## Confusion Matrix and Statistics
##
##           Reference
## Prediction  c0   c1
##         c0 131   27
##         c1  11   90
##
##                Accuracy : 0.8533
##                  95% CI : (0.8042, 0.894)
##     No Information Rate : 0.5483
##     P-Value [Acc > NIR] : < 2e-16
##
##                   Kappa : 0.7002
##
##  Mcnemar's Test P-Value : 0.01496
##
##             Sensitivity : 0.9225
##             Specificity : 0.7692
##          Pos Pred Value : 0.8291
##          Neg Pred Value : 0.8911
##              Prevalence : 0.5483
##          Detection Rate : 0.5058
##    Detection Prevalence : 0.6100
##       Balanced Accuracy : 0.8459
##
##        'Positive' Class : c0
```

Beyond the 7-Step R Pipeline   In the 7-step pipeline, we create a list of candidate models by different selections of predictors. There are other parameters, such as the kernel function, the value of C, that should be concerned in model selection. The R package `caret` can automate the process of cross-validation and facilitate the optimization of multiple parameters simultaneously. Below is an example

```
library(RCurl)
url <- paste0("https://raw.githubusercontent.com",
              "/analyticsbook/book/main/data/AD.csv")
AD <- read.csv(text=getURL(url))
str(AD)
#Train and Tune the SVM
n = dim(AD)[1]
n.train <- floor(0.8 * n)
idx.train <- sample(n, n.train)
AD[which(AD[,1]==0),1] = rep("Normal",length(which(AD[,1]==0)))
AD[which(AD[,1]==1),1] = rep("Diseased",length(which(AD[,1]==1)))
AD.train <- AD[idx.train,c(1:16)]
AD.test <- AD[-idx.train,c(1:16)]
trainX <- AD.train[,c(2:16)]
trainy= AD.train[,1]

## Setup for cross-validation:
# 10-fold cross validation
# do 5 repetitions of cv
# Use AUC to pick the best model

ctrl <- trainControl(method="repeatedcv",
                     repeats=1,
                     summaryFunction=twoClassSummary,
                     classProbs=TRUE)

# Use the expand.grid to specify the search space
grid <- expand.grid(sigma = c(0.002, 0.005, 0.01, 0.012, 0.015),
C = c(0.3,0.4,0.5,0.6)
)

# method: Radial kernel
# tuneLength: 9 values of the cost function
# preProc: Center and scale data
svm.tune <- train(x = trainX, y = trainy,
                  method = "svmRadial", tuneLength = 9,
                  preProc = c("center","scale"), metric="ROC",
                  tuneGrid = grid,
                  trControl=ctrl)

svm.tune
```

Then we can obtain the following results

```
## Support Vector Machines with Radial Basis Function Kernel
##
## 413 samples
##  15 predictor
##   2 classes: 'Diseased', 'Normal'
##
## Pre-processing: centered (15), scaled (15)
```

```
## Resampling: Cross-Validated (10 fold, repeated 1 times)
## Summary of sample sizes: 371, 372, 372, 371, 372, 372, ...
## Resampling results across tuning parameters:
##
##   sigma  C    ROC         Sens        Spec
##   0.002  0.3  0.8929523   0.9121053   0.5932900
##   0.002  0.4  0.8927130   0.8757895   0.6619048
##   0.002  0.5  0.8956402   0.8452632   0.7627706
##   0.002  0.6  0.8953759   0.8192105   0.7991342
##   0.005  0.3  0.8965129   0.8036842   0.8036797
##   0.005  0.4  0.8996565   0.7989474   0.8357143
##   0.005  0.5  0.9020830   0.7936842   0.8448052
##   0.005  0.6  0.9032422   0.7836842   0.8450216
##   0.010  0.3  0.9030514   0.7889474   0.8541126
##   0.010  0.4  0.9058248   0.7886842   0.8495671
##   0.010  0.5  0.9060999   0.8044737   0.8541126
##   0.010  0.6  0.9077848   0.8094737   0.8450216
##   0.012  0.3  0.9032308   0.7781579   0.8538961
##   0.012  0.4  0.9049043   0.7989474   0.8538961
##   0.012  0.5  0.9063505   0.8094737   0.8495671
##   0.012  0.6  0.9104511   0.8042105   0.8586580
##   0.015  0.3  0.9060412   0.7886842   0.8493506
##   0.015  0.4  0.9068165   0.8094737   0.8495671
##   0.015  0.5  0.9109051   0.8042105   0.8541126
##   0.015  0.6  0.9118615   0.8042105   0.8632035
##
## ROC was used to select the optimal model using  the largest
## value. The final values used for the model were
## sigma = 0.015 and C = 0.6.
```

Ensemble learning

Rationale and formulation

Ensemble learning is another example of how we design better learning algorithms. The random forest model is a particular case of **ensemble models**. An ensemble model consists of *K base models*, denoted as, $h_1, h_2, \ldots, h_K$. The algorithms to create ensemble models differ from each other in terms of the types of the base models, the way to create diversity in the base models, etc.

We have known the random forest model uses Bootstrap to create many datasets and builds a set of decision tree models. Some other ensemble learning methods, such as the **AdaBoost** model, also use decision tree as the base model. The two differ in the way to build a *diverse* set of base models. The framework of AdaBoost is illustrated in Figure 125. AdaBoost employs a sequential process to build its base models: it uses the original dataset (when the weights for the data points are equal) to build a decision tree; then it uses the decision tree to predict on the dataset, obtains the errors, and updates the weights of the data points[190]; then it builds another decision tree on the same dataset with the new weights, obtains the errors, and updates the weights of the data points again. The sequential process continues, until a given number of decision trees are built. This sequential process is designed for adaptability: later models focus more on the *hard* data points that present challenges for previous base models to achieve good prediction performance. Interested readers may find a formal presentation of the AdaBoost algorithm in the **Remarks** section.

[190] I.e., those data points that are wrongly classified will gain higher weights.

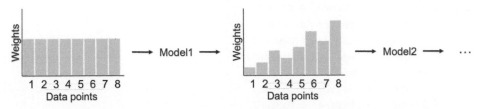

Figure 125: A general framework of AdaBoost

The ensemble learning is flexible, given that any model could be a base model. And there are a variety of ways to resample or perturb a dataset to create a diverse set of base models. Like SVM, the ensemble learning is another approach to have a built-in mechanism to reduce the risk of overfitting. Here, we provide a discussion of this built-in mechanism using the framework proposed by Dietterich[191], where three perspectives (statistical, computational, and representational) were used to explain why ensemble methods could lead to robust performance. Each perspective is described in details below.

[191] Dietterich, T.G., *Ensemble methods in machine learning*, Multiple Classifier Systems, Springer, 2000.

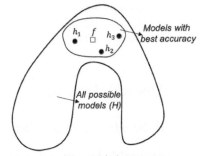

Figure 126: Ensemble learning approximates the true model with a combination of good models (statistical perspective)

Statistical perspective   The statistical reason is illustrated in Figure 126. $\mathcal{H}$ is the model space where a learning algorithm searches for the best model guided by the training data. A model corresponds to a *point* in Figure 126, e.g., the point labelled as f is the true model. When the data is limited and the best models are multiple, the problem is a statistical one and we need to make an optimal decision despite the uncertainty. This is illustrated by the inner circle in Figure 126. By building an ensemble of multiple base models, e.g., the h_1, h_2, and h_3 in Figure 126, the average of the models is a good approximation to the true model f. This combined solution, comparing with other models that only identify one best model, has less variance, and therefore, could be more robust.

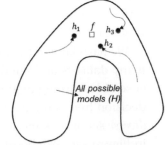

Figure 127: Ensemble learning provides a robust coverage of the true model (computational perspective)

Computational perspective   A computational perspective is shown in Figure 127. This perspective concerns the way we build base models. Often greedy approaches such as the recursive splitting procedure are used to solve optimization problems in training machine learning models. This is optimal only in a *local* sense[192]. As a remedy to this problem, the ensemble learning initializes the learning algorithm (that is greedy and heuristic) from multiple locations in $\mathcal{H}$, i.e., as shown in Figure 127, three models are identified by the same algorithm that starts from different initial points. Exploring multiple trajectories help us find a robust coverage of the true model f.

Representational perspective   Due to the size of the dataset or the limitations of a model, sometimes the model space $\mathcal{H}$ does not cover the true model, i.e., in Figure 128 the true model is outside the region of $\mathcal{H}$. This is not uncommon in real-world problems, for example, linear models cannot learn nonlinear patterns, or decision trees have difficulty in learning linear patterns. Using multiple base models may provide an approximation of the true model that is outside $\mathcal{H}$, as shown in Figure 128.

Analysis of the decision tree, random forests, and AdaBoost

The three models are analyzed using the three perspectives. Results are shown in Table 31. In-depth discussions are provided in the following.

Single decision tree   A single decision tree lacks the capability to overcome overfitting in terms of each of the three perspectives. From the statistical perspective, a decision tree algorithm constructs each node using the maximum information gain *at that particular node only*; thus, random errors in data may mislead subsequent splits. On the other hand, when the training dataset is limited, many models may perform equally well, since there are not enough data to distinguish these models. This results in a large *inner circle* as shown in Figure 126. With the true model f hidden in a large area in $\mathcal{H}$, and the sensitivity of the learning algorithm to random noises in data (an issue from the computational perspective), the learning algorithm may end up with a model far away from the true model f.

Perspectives	DT	RF	AdaBoost
Statistical	No	Yes	No
Computational	No	Yes	Yes
Representational	No	No	Yes

From the representational perspective, there are also limitations of the decision tree model; i.e., in **Chapter 2** we have shown that the decision tree model has difficulty in modeling linear patterns in the data.

[192] E.g., to grow a decision tree, at each node, the node is split according to the maximum information gain *at this particular node*. To grow a decision tree model, a sequence of splits is needed. Optimization of all the splits *simultaneously* leads to a *global* optimal solution, but it is a *NP-hard* problem that is not solved yet. Optimization of each split is more practical, only we know that the local optimal solution may result in suboptimal situations for further splitting of descendant nodes.

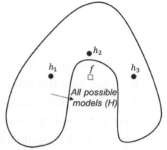

Figure 128: Ensemble learning approximates the true model with a combination of good models (representational perspective)

Table 31: Analysis of the decision tree (DT), random forests (RF), and AdaBoost using the three perspectives

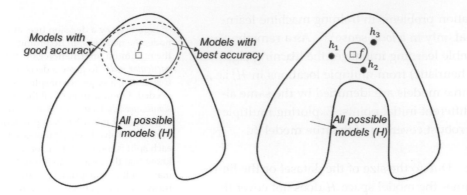

Figure 129: Analysis of the random forest in terms of the statistical (left), computational (middle), and representational (right) perspectives

Random forests   From the statistical perspective, the random forest model is a good ensemble learning model. As shown in Figure 129 (left), the way the random forest model grows the base models is to construct the *circle* of dotted line. Models located in this circle of dotted line have reasonably good accuracy. These models may not be the best models with great accuracy, they do provide a good coverage/approximation of the true model.

Note that, if we could directly build a model that is close to f, or build many best models that are located in the circle of dotted line, that would be ideal. However, both tasks are challenging. Comparing with these ideal goals, the random forest model is more pragmatic. It cleverly uses *simple*[193] techniques of *randomness*, i.e., the Bootstrap and the random selection of variables, that are robust, effective, and easy to implement. It grows a set of models that are not the best, but good models. Most importantly, these good models complement each other[194].

Random forest model can also address the computational issue. As shown in Figure 129 (middle), while the circle of solid line (i.e., that represents the space of best models) is computationally difficult to reach, averaging multiple models could provide a good approximation.

It seems that the random forest models do not actively solve the representational issue. If the true model f lies outside $\mathcal{H}$, as shown in Figure 129 (right), averaging multiple models won't necessarily approximate the true model.

AdaBoost   Similar to random forest, AdaBoost solves the computational issue by generating many base models. The difference is that, AdaBoost actively solves the representational issue, i.e., it tries to do better on the *hard* data points where the previous base models fail to predict correctly. For each base model in AdaBoost, the training dataset is not resampled by Bootstrap, but weighted based on the

[193] As we have seen, *Simple* is a complex word.

[194] In practice, the challenge to grow a set of *best* models is that it usually ends up with these *best models* more or less being the same.

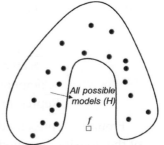

Figure 130: Analysis of the AdaBoost in terms of the representational perspective

error rates from previous base models, i.e., data points that are difficult to be correctly predicted by the previous models are given more weights in the new training dataset for the subsequent base model. Figure 130 shows this sequential learning process helps AdaBoost identify more models around the true model, and put more weight to the models that are closer to the true model.

But AdaBoost is not as good as random forest in terms of addressing the statistical issue. As AdaBoost aggressively solves the representational issue and allows its base models to be impacted by some *hard* data points[195], it is more likely to overfit, and may be less stable than the random forest models that place more emphasis on addressing the statistical issue.

[195] This is a common root cause for a model to overfit the training data, if the model tries *too hard* on a particular training data.

R Lab

We use the AD dataset to study decision tree (`rpart` package), random forests (`randomForest` package), and AdaBoost (`gbm` package).

First, we evaluate the overall performance of the three models. Results are shown in Figure 131, produced by the following R code.

```
theme_set(theme_gray(base_size = 15))
library(randomForest)
library(gbm)
library(rpart)
library(dplyr)
library(RCurl)
url <- paste0("https://raw.githubusercontent.com",
             "/analyticsbook/book/main/data/AD.csv")
data <- read.csv(text=getURL(url))

rm_indx <- which(colnames(data) %in% c("ID", "TOTAL13",
                                       "MMSCORE"))
data <- data[, -rm_indx]
data$DX_bl <- as.factor(data$DX_bl)

set.seed(1)

err.mat <- NULL
for (K in c(0.2, 0.3, 0.4, 0.5, 0.6, 0.7)) {

testing.indices <- NULL
for (i in 1:50) {
testing.indices <- rbind(testing.indices, sample(nrow(data),
                    floor((1 - K) * nrow(data))))
}

for (i in 1:nrow(testing.indices)) {
```

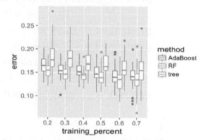

Figure 131: Boxplots of the classification error rates for single decision tree, random forest, and AdaBoost

```
testing.ix <- testing.indices[i, ]
target.testing <- data$DX_bl[testing.ix]

tree <- rpart(DX_bl ~ ., data[-testing.ix, ])
pred <- predict(tree, data[testing.ix, ], type = "class")
error <- length(which(as.character(pred) !=
                target.testing))/length(target.testing)
err.mat <- rbind(err.mat, c("tree", K, error))

rf <- randomForest(DX_bl ~ ., data[-testing.ix, ])
pred <- predict(rf, data[testing.ix, ])
error <- length(which(as.character(pred) !=
                target.testing))/length(target.testing)
err.mat <- rbind(err.mat, c("RF", K, error))

data1 <- data
data1$DX_bl <- as.numeric(as.character(data1$DX_bl))
boost <- gbm(DX_bl ~ ., data = data1[-testing.ix, ],
            dist = "adaboost",interaction.depth = 6,
            n.tree = 2000)  #cv.folds = 5,
# best.iter <- gbm.perf(boost,method='cv')
pred <- predict(boost, data1[testing.ix, ], n.tree = 2000,
                type = "response")  # best.iter n.tree = 400,
pred[pred > 0.5] <- 1
pred[pred <= 0.5] <- 0
error <- length(which(as.character(pred) !=
                    target.testing))/length(target.testing)
err.mat <- rbind(err.mat, c("AdaBoost", K, error))
  }
}
err.mat <- as.data.frame(err.mat)
colnames(err.mat) <- c("method", "training_percent", "error")
err.mat <- err.mat %>% mutate(training_percent =
    as.numeric(as.character(training_percent)), error =
    as.numeric(as.character(error)))

ggplot() + geom_boxplot(data = err.mat %>%
        mutate(training_percent = as.factor(training_percent)),
        aes(y = error, x = training_percent,
            color = method)) + geom_point(size = 3)
```

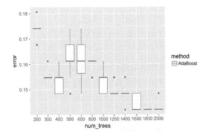

Figure 131 shows that the decision tree is less accurate than the other two ensemble methods. The random forest has lower error rates than AdaBoost in general. As the training data size increases, the gap between random forest and AdaBoost decreases. This may indicate that when the training data size is small, the random forest is more stable due to its advantage of addressing the statistical issue. Overall, all models become better as the percentage of the training data increases.

Figure 132: Boxplots of the classification error rates for AdaBoost with a different number of trees

We adjust the number of trees in AdaBoost and show the results in Figure 132. It can be seen that the error rates first go down as the number of trees increases to 400. Then the error rates increase, and decrease again. The unstable relationship between the error rates with the number of trees of AdaBoost indicates that AdaBoost is impacted by some particularity of the dataset and seems less robust than random forest.

```r
err.mat <- NULL
set.seed(1)
for (i in 1:nrow(testing.indices)) {
  data1 <- data
  data1$DX_bl <- as.numeric(as.character(data1$DX_bl))
  ntree.v <- c(200, 300, 400, 500, 600, 800, 1000, 1200,
               1400, 1600, 1800, 2000)
  for (j in ntree.v) {
    boost <- gbm(DX_bl ~ ., data = data1[-testing.ix, ],
                 dist = "adaboost", interaction.depth = 6,
                 n.tree = j)
    # best.iter <- gbm.perf(boost,method='cv')
    pred <- predict(boost, data1[testing.ix, ], n.tree = j,
                    type = "response")
    pred[pred > 0.5] <- 1
    pred[pred <= 0.5] <- 0
    error <- length(which(as.character(pred) !=
                    target.testing))/length(target.testing)
    err.mat <- rbind(err.mat, c("AdaBoost", j, error))
  }
}
err.mat <- as.data.frame(err.mat)
colnames(err.mat) <- c("method", "num_trees", "error")
err.mat <- err.mat %>%
  mutate(num_trees = as.numeric(as.character(num_trees)),
         error = as.numeric(as.character(error)))

ggplot() + geom_boxplot(data = err.mat %>%
           mutate(num_trees = as.factor(num_trees)),
           aes(y = error, x = num_trees, color = method)) +
               geom_point(size = 3)
```

We repeat the experiment on random forest and show the result in Figure 133. Similar to AdaBoost, when the number of trees is small, the random forest has higher error rates. Then, the error rates decrease as more trees are added. And the error rates become stable when more trees are added. The random forest handles the statistical issue better than the AdaBoost.

```r
err.mat <- NULL
set.seed(1)
for (i in 1:nrow(testing.indices)) {
```

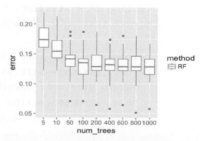

Figure 133: Boxplots of the classification error rates for random forests with a different number of trees

```
testing.ix <- testing.indices[i, ]
target.testing <- data$DX_bl[testing.ix]

ntree.v <- c(5, 10, 50, 100, 200, 400, 600, 800, 1000)
for (j in ntree.v) {
rf <- randomForest(DX_bl ~ ., data[-testing.ix, ], ntree = j)
pred <- predict(rf, data[testing.ix, ])
error <- length(which(as.character(pred) !=
                            target.testing))/length(target.testing)
err.mat <- rbind(err.mat, c("RF", j, error))
}
}
err.mat <- as.data.frame(err.mat)
colnames(err.mat) <- c("method", "num_trees", "error")
err.mat <- err.mat %>% mutate(num_trees =
                        as.numeric(as.character(num_trees)),
error = as.numeric(as.character(error)))

ggplot() + geom_boxplot(data =
            err.mat %>% mutate(num_trees = as.factor(num_trees)),
            aes(y = error, x = num_trees, color = method)) +
            geom_point(size = 3)
```

Building on the result shown in Figure 133, we pursue a further study of the behavior of random forest. Recall that, in random forest, there are two approaches to increase diversity, one is to Bootstrap samples for each tree, while another is to conduct random feature selection for splitting each node.

First, we investigate the effectiveness of the use of Bootstrap. We change the sampling strategy from *sampling with replacement* to *sampling without replacement* and change the sampling size[196] from 10% to 100%. The number of features tested at each node is kept at the default value, i.e., $\sqrt{p}$, where p is the number of features. Figure 134 shows that the increased sample size has an impact on the error rates.

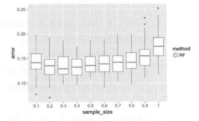

Figure 134: Boxplots of the classification error rates for random forest with a different sample sizes

[196] The sampling size is the sample size of the Bootstrapped dataset.

```
err.mat <- NULL
set.seed(1)
for (i in 1:nrow(testing.indices)) {
  testing.ix <- testing.indices[i, ]
  target.testing <- data$DX_bl[testing.ix]

  sample.size.v <- seq(0.1, 1, by = 0.1)
  for (j in sample.size.v) {
    sample.size <- floor(nrow(data[-testing.ix, ]) * j)
    rf <- randomForest(DX_bl ~ ., data[-testing.ix, ],
                    sampsize = sample.size,
                    replace = FALSE)
    pred <- predict(rf, data[testing.ix, ])
    error <- length(which(as.character(pred) !=
```

```
                 target.testing))/length(target.testing)
      err.mat <- rbind(err.mat, c("RF", j, error))
    }
}
err.mat <- as.data.frame(err.mat)
colnames(err.mat) <- c("method", "sample_size", "error")
err.mat <- err.mat %>% mutate(sample_size =
               as.numeric(as.character(sample_size)),
               error = as.numeric(as.character(error)))
ggplot() + geom_boxplot(data = err.mat %>%
             mutate(sample_size = as.factor(sample_size)),
             aes(y = error, x = sample_size,color = method)) +
          geom_point(size = 3)
```

We then investigate the effectiveness of using random selection of features for node splitting. We fix the sampling size to be the same size as the original dataset, and change the number of features to be selected. Results are shown in Figure 135. When the number of features reaches 11, the error rate starts to increase. This is probably because of the loss of the diversity of the trees, i.e., the more features to be used, the less randomness is introduced into the trees.

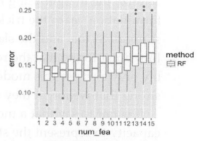

Figure 135: Boxplots of the classification error rates for random forest with a different number of features

```
err.mat <- NULL
set.seed(1)
for (i in 1:nrow(testing.indices)) {
  testing.ix <- testing.indices[1, ]
  target.testing <- data$DX_bl[testing.ix]

  num.fea.v <- 1:(ncol(data) - 1)
  for (j in num.fea.v) {
    sample.size <- nrow(data[-testing.ix, ])
    rf <- randomForest(DX_bl ~ ., data[-testing.ix, ],
                      mtry = j, sampsize = sample.size,
                      replace = FALSE)
    pred <- predict(rf, data[testing.ix, ])
    error <- length(which(as.character(pred) !=
                 target.testing))/length(target.testing)
    err.mat <- rbind(err.mat, c("RF", j, error))
  }
}
err.mat <- as.data.frame(err.mat)
colnames(err.mat) <- c("method", "num_fea", "error")
err.mat <- err.mat %>% mutate(num_fea
                  = as.numeric(as.character(num_fea)),
                 error = as.numeric(as.character(error)))

ggplot() + geom_boxplot(data =
           err.mat %>% mutate(num_fea = as.factor(num_fea)),
            aes(y = error, x = num_fea, color = method)) +
          geom_point(size = 3)
```

Remarks

Is SVM a more complex model?

In the preface of his seminar book[197], Vladimir Vapnik wrote that *"...during the last few years at different computer science conferences, I heard reiteration of the following claim: 'Complex theories do not work, simple algorithms do'...this is not true...Nothing is more practical than a good theory...".* He created the concept of *VC dimension* to specifically characterize his concept of the complexity of a model.

A model is often *perceived* to be complex. The SVM model looks more complex than the linear regression model. It asks us to characterize the margin using model parameters, write the optimization formulation, learn the trick of kernel function, and understand the support vectors and the slack variables for the nonseparable case. But, don't forget that the reason for a model to look simple is probably only because this model may presuppose stronger conditions, too strong that we forget they are assumptions.

It is fair to say that a model is more complex if it provides more capacity to represent the statistical phenomena in the training data. In other words, a more complex model is more flexible to respond to subtle patterns in the data by adjusting itself. In this sense, SVM with kernel functions is a complex model since it can model nonlinearity in the data. But on the other hand, comparing the SVM model with other linear models as shown in Figure 136, it is hard to tell that the SVM model is simpler, but it is clear that it is more stubborn; because of its pursuit of maximum margin, it ends up with one model only. If you are looking for an example of an idea that is radical and conservative, flexible and disciplined, this is it.

[197] Vapnik, V., *The Nature of Statistical Learning Theory*, Springer, 2000.

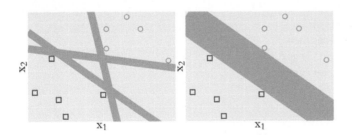

Figure 136: (Left) some other linear models; (b) the SVM model

Is SVM a neural network model?

Another interesting fact about SVM is that, when it was developed, it was named "support vector network"[198]. In other words, it has a connection with the artificial neural network that will be discussed in **Chapter 10**. This is revealed in Figure 137. Readers who know neural

[198] Cortes, C. and Vapnik, V., *Support-vector networks*, Machine Learning, Volume 20, Issue 3, Pages 273–297, 1995.

network models are encouraged to write up the mathematical model of the SVM model following the neural network format as shown in Figure 137.

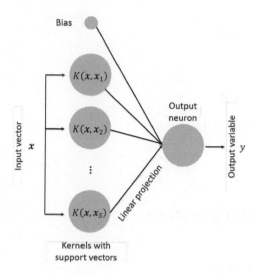

Figure 137: SVM as a neural network model

Derivation of the margin

Consider any two points on the two margins, e.g., the x_A and x_B in Figure 138. The *margin width* is equal to the projection of the vector $\overrightarrow{AB} = x_B - x_A$ on the direction w, which is

$$\text{margin} = \frac{(x_B - x_A) \cdot \vec{w}}{\|w\|}. \tag{85}$$

It is known that

$$w^T x_B + b = 1,$$

and

$$w^T x_A + b = -1.$$

Thus, Eq. 85 is rewritten as

$$\text{margin} = \frac{2}{\|w\|}. \tag{86}$$

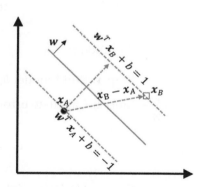

Figure 138: Illustration of how to derive the margin

Why the nonzero α_n are the support vectors

Theoretically, to understand why the nonzero α_n are the support vectors, we can use the *Karush–Kuhn–Tucker (KKT) conditions*[199]. Based on the *complementary slackness* as one of the KKT conditions, the following equations must hold

[199] Bertsekas, D., *Nonlinear Programming: 3rd Edition*, Athena Scientific, 2016.

$$\alpha_n \left[y_n \left(w^T x_n + b \right) - 1 \right] = 0, \text{ for } n = 1, 2, \ldots, N.$$

Thus, for any data point x_n, it is either

$$\alpha_n = 0, \text{ and } y_n \left(w^T x_n + b \right) - 1 \neq 0;$$

or

$$\alpha_n \neq 0, \text{ and } y_n \left(w^T x_n + b \right) - 1 = 0.$$

Revisiting Eq. 63 or Figure 118, we know that only the support vectors have $\alpha_n \neq 0$ and $y_n \left(w^T x_n + b \right) - 1 = 0$.

AdaBoost algorithm

The specifics of the AdaBoost algorithm shown in Figure 125 are described below.

- Input: N data points, $(x_1, y_1), (x_2, y_2), \ldots, (x_N, y_N)$.

- Initialization: Initialize equal weights for all data points

$$w_0 = \left(\frac{1}{N}, \ldots, \frac{1}{N} \right).$$

- At iteration t:

  - Step 1: Build model h_t on the dataset with weights w_{t-1}.

  - Step 2: Calculate errors using h_t

$$\epsilon_t = \sum_{n=1}^{N} w_{t,n} \left\{ h_t \left(x_n \right) \neq y_n \right\}.$$

  - Step 3: Update weights of the data points

$$w_{t+1,i} = \frac{w_{t,i}}{Z_t} \times \begin{cases} e^{-\alpha_t} \text{ if } h_t \left(x_n \right) = y_n \\ e^{\alpha_t} \text{ if } h_t \left(x_n \right) \neq y_n. \end{cases}$$

Here,

Z_t is a normalization factor so that $\sum_{n=1}^{N} w_{t+1,n} = 1,$

and

$$\alpha_t = \frac{1}{2} \ln \left(\frac{1 - \epsilon_t}{\epsilon_t} \right).$$

- Iterations: Repeat Step 1 to Step 3 for T times, to get $h_1, h_2, h_3, \ldots, h_T$.

- Output:

$$H(x) = \text{sign}\left(\sum_{t=1}^{T} \alpha_t h_t(x)\right).$$

When all the base models are trained, the aggregation of these models in predicting on a data instance x is a weighted sum of base models

$$h(x) = \sum_i \gamma_i h_i(x),$$

where the weight γ_i is proportional to the accuracy of $h_i(x)$ on the training dataset.

Exercises

1. To build a linear SVM on the data shown in Figure 139, how many support vectors are needed (use visual inspection)?

2. Let's consider the dataset in Table 32. Please (a) draw scatterplots and identify the support vectors if you'd like to build a linear SVM classifier; (b) manually derive the alpha values (i.e., the α_i) for the support vectors and the offset parameter b; (c) derive the weight vector (i.e., the $\hat{w}$) of the SVM model; and (d) predict on the new dataset and fill in the column of y in Table 33.

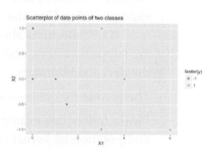

Figure 139: How many support vectors are needed?

Table 32: Dataset for building a SVM model in Q2

ID	x_1	x_2	x_3	y
1	4	1	1	1
2	4	−1	0	1
3	8	2	1	1
4	−2.5	0	0	−1
5	0	1	1	−1
6	−0.3	−1	0	−1
7	2.5	−1	1	−1
8	−1	1	0	−1

ID	x_1	x_2	x_3	y
9	5.4	1.2	2	
10	1.5	−2	3	
11	−3.4	1	−2	
12	−2.2	−1	−4	

Table 33: Test data points for the SVM model in Q2

3. Follow up on the dataset used in Q2. Use the R pipeline for SVM on this data. Compare the alpha values (i.e., the α_i), the offset parameter b, and the weight vector (i.e., the $\hat{w}$) from R and the result by your manual calculation in Q2.

4. Modify the R pipeline for Bootstrap and incorporate the `glm` package to write your own version of ensemble learning that ensembles a set of logistic regression models. Test it using the same data that has been used in the R lab for logistic regression models.

5. Use the dataset `PimaIndiansDiabetes2` in the `mlbench` R package, run the R SVM pipeline on it, and summarize your findings.

6. Use R to generate a dataset with two classes as shown in Figure 140. Then, run SVM model with a properly selected kernel function on this dataset.

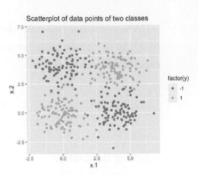

Figure 140: A dataset with two classes

7. Follow up on the dataset generated in Q6. Try visualizing the decision boundaries by different kernel functions such as linear, Laplace, Gaussian, and polynomial kernel functions. Below is one example using Gaussian kernel with its bandiwidth parameter $\gamma = 0.2$.[200] Result is shown in Figure 141. The blackened points are support vectors, and the contour reflects the characteristics of the decision boundary.

The R code for generating Figure 141 is shown below.

Please follow this example and visualize linear, Laplace, Gaussian, and polynomial kernel functions with different parameter values.

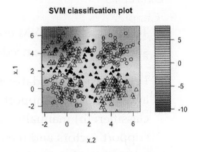

Figure 141: Visualization of the decision boundary of an SVM model with Gaussian kernel

[200] In the following R code, the bandiwidth parameter is specified as `sigma=0.2` .

```
require( 'kernlab' )
rbf.svm <- ksvm(y ~ ., data=data, type='C-svc', kernel='rbfdot',
                kpar=list(sigma=0.2), C=100, scale=c())
plot(rbf.svm, data=data)
```

Chapter 8: Scalability
LASSO & PCA

Overview

Chapter 8 is about *Scalability*. **LASSO** and **PCA** will be introduced. LASSO stands for the **least absolute shrinkage and selection operator,** which is a representative method for *feature selection*. PCA stands for the **principal component analysis**, which is a representative method for *dimension reduction*. Both methods can reduce the dimensionality of a dataset but follow different styles. LASSO, as a feature selection method, focuses on deletion of irrelevant or redundant features. PCA, as a dimension reduction method, combines the features into a smaller number of aggregated components (a.k.a., the new features). A remarkable difference between the two approaches is that, while both create a dataset with a smaller dimensionality, in PCA the original features are used to derive the new features[201].

[201] As a result, no feature is discarded.

LASSO

Rationale and formulation

Two points determine a line, as shown in Figure 142. It shares the same geometric form as a linear regression model, but it is a deterministic geometric pattern and has nothing to do with *error*.

With one more data point, magic happens: as shown in Figure 143, now we can estimate the *residuals* and study the systematic patterns of *error*. The line in Figure 143 becomes a statistical model.

The two lines in Figures 142 and 143, one is a deterministic pattern, while another is a statistical model, are like *homonym*. The different meanings share the same form of their signifier (e.g., like the word *bass* that means a certain sound that is low and deep, or a type of fish).

The *error* is a defining component of a statistical model. It models the *noise* in the data. In an application context, understanding the

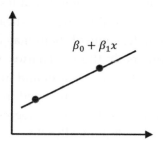

Figure 142: A line is not a model

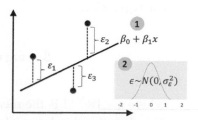

Figure 143: Revisit the linear regression model

noise and knowing how much proportion the noise contributes to the total variation of the dataset is important knowledge. And, to derive the *p-values* of the regression coefficients, we need the noise so that we can compare the strength of the estimated coefficients with the noise to evaluate if the estimated coefficients are significantly different from random manifestation (i.e., if we cannot model the noise, then we have no basis to define what is random manifestation.).

To model a linear regression model, we need enough data points to estimate the error. For the simple example when there is only one predictor x, as shown in Figures 142 and 143, we would need at least 3 data points to estimate the error[202]. This is just enough to solve for the 2 regression coefficients. Consider a problem with 10 variables, what is the minimum number of data points needed to enable the estimation of error[203]?

From the examples aforementioned, we could deduce that the number of data points, i.e., denoted as N, needs to be larger than the number of variables, i.e., denoted as p. This is barely a minimum requirement of linear regression, as we haven't asked how many data points are needed to ensure high-quality estimation of the parameters. In classic settings in statistics, N is assumed to be much larger than p in order to prove asymptotics—a common approach to prove a statistical model is valid. Practically, linear regression model finds difficulty in applications where the ratio N/p is small. In recent years, there are applications where the number of data points is even smaller than the number of variables, i.e., commonly referred to as $N < p$ problems.

When increasing N is not always a feasible option, reducing p is a necessity. Some variables may be irrelevant or simply noise. Even if all variables are statistically informative, when considered as a whole, some of them may be redundant, and some are weaker than others. In those scenarios, there is room for us to wriggle with the problematic dataset and improve on the ratio N/p by reducing p.

LASSO was invented in 1996 to sparsify the linear regression model and allow the regression model to select significant predictors automatically[204].

Remember that, to estimate β, the least squares estimation of linear regression is

$$\hat{\beta} = \arg\min_{\beta} (y - X\beta)^T (y - X\beta), \tag{87}$$

where $y \in \mathbb{R}^{N \times 1}$ is the measurement vector of the outcome variable, $X \in \mathbb{R}^{N \times p}$ is the data matrix of the N measurement vectors of the p predictors, $\beta \in \mathbb{R}^{p \times 1}$ is the regression coefficient vector[205].

[202] While this is obvious from Figures 142 and 143, we can also obtain the conclusion by derivation. I.e., given two data points, (x_1, y_1) and (x_2, y_2) we could write two equations, $y_1 = \beta_0 + \beta_1 x_1$ and $y_2 = \beta_0 + \beta_1 x_2$.

[203] The answer is 12. Suppose that we only have 11 data points. The regression *line* is defined by 11 regression coefficients. For each data point, we can write up an equation. Thus, 11 data points are just enough to estimate the 11 regression coefficients, leaving no room for estimating errors.

[204] Tibshirani, R. *Regression shrinkage and selection via the Lasso*, Journal of the Royal Statistical Society (Series B), Volume 58, Issue 1, Pages 267-288, 1996.

[205] Here, we assume that the data is normalized/standardized and no intercept coefficient β_0 is needed. Normalization means $\sum_{n=1}^{N} x_{nj}/N = 0$, $\sum_{n=1}^{N} x_{ij}^2/N = 1$ for $j = 1, 2, \ldots, p$ and $\sum_{n=1}^{N} y_n/N = 0$. Normalization is a common practice, and some R packages automatically normalize the data as a default preprocessing step before the application of a model.

The formulation of LASSO is

$$\hat{\beta} = \arg\min_{\beta} \left\{ \underbrace{(y - X\beta)^T(y - X\beta)}_{\text{Least squares}} + \underbrace{\lambda\|\beta\|_1}_{L_1 \text{ norm penalty}} \right\} \quad (88)$$

where $\|\beta\|_1 = \sum_{i=1}^{p} |\beta_i|$. The parameter, λ, is called the **penalty parameter** that is specified by user of LASSO. The larger the parameter λ, the more zeros in $\hat{\beta}$.

It could be seen that LASSO embodies two components in its formulation. The 1st term is the least squares loss function inherited from linear regression that is used to measure the goodness-of-fit of the model. The 2nd term is the sum of absolute values of elements in β that is called the L_1 norm penalty. It measures the *model complexity*, i.e., smaller $\|\beta\|_1$ tends to create more zeros in β, leading to a simpler model. In practice, by tuning the parameter λ, we hope to find the best model with an optimal balance between model fit and model complexity.

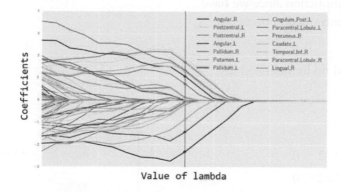

Figure 144: Path solution trajectory of the coefficients; each curve corresponds to a regression coefficient

As shown in Figure 144, LASSO can generate a **path solution trajectory** that visualizes the solutions of β for a continuum of values of λ. Model selection criteria such as the Akaike Information Criteria (AIC) or cross-validation can be used to identify the best λ that would help us find the final model, i.e., as the vertical line shown in Figure 144. When many variables are deleted from the model, the dimensionality of the model is reduced, and N/p is increased.

The shooting algorithm

We introduce the **shooting algorithm** to solve for the optimization problem shown in Eq. 88. Let's consider a simple example when there is only one predictor x. The objective function in Eq. 88 could be rewritten as

$$l(\beta) = (y - X\beta)^T(y - X\beta) + \lambda|\beta|. \qquad (89)$$

To solve Eq. 89, we take the differential of $l(\beta)$ and put it equal to zero

$$\frac{\partial l(\beta)}{\partial \beta} = 0. \qquad (90)$$

A complication of this differential operation is that the L_1-norm term, $|\beta|$, has no gradient when $\beta = 0$. There are three scenarios:

- If $\beta > 0$, then $\frac{\partial L(\beta)}{\partial \beta} = 2\beta - 2X^Ty + \lambda$. Based on Eq. 90, we can estimate β as $\hat{\beta} = X^Ty - \lambda/2$. Note that this estimate of β may turn out to be negative. If that happens, it would be a contradiction since we have assumed $\beta > 0$ to derive the result. This contradiction points to the only possibility that $\beta = 0$.

- If $\beta < 0$, then $\frac{\partial L(\beta)}{\partial \beta} = 2\beta - 2X^Ty - \lambda$. Based on Eq. 90, we have $\beta = X^Ty + \lambda/2$. Note that this estimate of β may turn out to be positive. If that happens, it would be a contradiction since we have assumed $\beta < 0$ to derive the result. This contradiction points to the only possibility that $\beta = 0$.

- If $\beta = 0$, then we have had the solution and no longer need to derive the gradient.

In summary, the solution of β is

$$\hat{\beta} = \begin{cases} X^Ty - \lambda/2, & if\ X^Ty - \lambda/2 > 0 \\ X^Ty + \lambda/2, & if\ X^Ty + \lambda/2 < 0 \\ 0, & if\ \lambda/2 \geq |X^Ty|. \end{cases} \qquad (91)$$

Now let's consider the general case as shown in Eq. 88. Figure 145 illustrates the basic idea: to apply the conclusion (with a slight variation) we have obtained in Eq. 91 to solve Eq. 88. Each iteration solves for one regression coefficient, assuming that all the other coefficients are fixed (i.e., to their latest values).

In each iteration, we solve a similar problem with the one-predictor special problem shown in Eq. 89. For instance, denote $\beta_j^{(t)}$ as the estimate of β_j in the tth iteration. If we fix the other regression coefficients as their latest estimates, we can rewrite Eq. 88 as a function of $\beta_j^{(t)}$ only

$$l(\beta_j^{(t)}) = \left(y_j^{(t)} - X_{(:,j)}\beta_j^{(t)}\right)^T \left(y_j^{(t)} - X_{(:,j)}\beta_j^{(t)}\right) + \lambda|\beta_j^{(t)}|, \qquad (92)$$

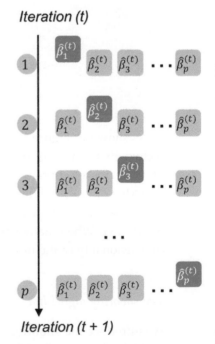

Figure 145: The shooting algorithm iterates through the coefficients

where $X_{(:,j)}$ is the jth column of the matrix X, and

$$y_j^{(t)} = y - \sum_{k \neq j} X_{(:,k)} \hat{\beta}_k^{(t)}. \tag{93}$$

Eq. 92 has the same structure as Eq. 89. We can readily apply the conclusion in Eq. 91 here and obtain

$$\hat{\beta}_j^{(t)} = \begin{cases} q_j^{(t)} - \lambda/2, & if \, q_j^{(t)} - \lambda/2 > 0 \\ q_j^{(t)} + \lambda/2, & if \, q_j^{(t)} + \lambda/2 < 0 \\ 0, & if \, \lambda/2 \geq |q_j^{(t)}|, \end{cases} \tag{94}$$

where

$$q_j^{(t)} = X_{(:,j)}^T y_j^{(t)}. \tag{95}$$

A small data example

x_1	x_2	y
-0.707	0	-0.77
0	0.707	0.15
0.707	-0.707	0.62

Table 34: A dataset example for LASSO

Consider a dataset example as shown in Table 34.[206]
In matrix form, the dataset is rewritten as

[206] To generate this dataset, we sampled the values of y using the model

$y = 0.8x_1 + \varepsilon$, where $\varepsilon \sim N(0, 0.5)$.

Only x_1 is important.

$$y = \begin{bmatrix} -0.77 \\ 0.15 \\ 0.62 \end{bmatrix}, \, X = \begin{bmatrix} -0.707 & 0 \\ 0 & 0.707 \\ 0.707 & -0.707 \end{bmatrix}.$$

Now let's implement the *Shooting algorithm* on this data. The objective function of LASSO on this case is

$$\sum_{n=1}^{3} [y_n - (\beta_1 x_{n,1} + \beta_2 x_{n,2})]^2 + \lambda(|\beta_1| + |\beta_2|).$$

Suppose that $\lambda = 1$, and we initiate the regression coefficients as $\hat{\beta}_1^{(0)} = 0$ and $\hat{\beta}_2^{(0)} = 1$.

To update $\hat{\beta}_1^{(1)}$, based on Eq. 94, we first calculate $y_1^{(1)}$ using Eq. 93

$$y_1^{(1)} = y - X_{(:,2)}\hat{\beta}_2^{(0)} = \begin{bmatrix} -0.7700 \\ -0.557 \\ 1.3270 \end{bmatrix}.$$

Then we calculate $q_1^{(1)}$ using Eq. 95

$$q_1^{(1)} = X_{(:,1)}^T y_1^{(1)} = 1.7654.$$

As
$$q_1^{(1)} - \lambda/2 > 0,$$
based on Eq. 94 we know that
$$\hat{\beta}_1^{(1)} = q_1^{(1)} - \lambda/2 = 1.2654.$$

Then we update $\hat{\beta}_2^{(1)}$. We can obtain that
$$\boldsymbol{y}_2^{(1)} = \boldsymbol{y} - \boldsymbol{X}_{(:,1)}\hat{\beta}_1^{(1)} = \begin{bmatrix} 0.1876 \\ 0.1500 \\ -0.2746 \end{bmatrix}.$$

And we can get
$$q_2^{(1)} = \boldsymbol{X}_{(:,2)}^T \boldsymbol{y}_2^{(1)} = 0.3002.$$

As
$$\lambda/2 \geq |q_2^{(1)}|,$$
we know that
$$\hat{\beta}_2^{(1)} = 0.$$

Thus, with only one iteration, the *Shooting algorithm* identified the irrelevant variable.

R Lab

The 7-Step R Pipeline **Step 1** and **Step 2** get dataset into R and organize the dataset in required format.

```
# Step 1 -> Read data into R workstation
#### Read data from a CSV file
#### Example: Alzheimer's Disease
# RCurl is the R package to read csv file using a link
library(RCurl)
url <- paste0("https://raw.githubusercontent.com",
              "/analyticsbook/book/main/data/AD_hd.csv")
AD <- read.csv(text=getURL(url))
str(AD)
```

```
# Step 2 -> Data preprocessing
# Create your X matrix (predictors) and Y
# vector (outcome variable)
X <- AD[,-c(1:4)]
Y <- AD$MMSCORE

# Then, we integrate everything into a data frame
data <- data.frame(Y,X)
names(data)[1] = c("MMSCORE")

# Create a training data
```

```
train.ix <- sample(nrow(data),floor( nrow(data)) * 4 / 5 )
data.train <- data[train.ix,]
# Create a testing data
data.test <- data[-train.ix,]

# as.matrix is used here, because the package
# glmnet requires this data format.
trainX <- as.matrix(data.train[,-1])
testX <- as.matrix(data.test[,-1])
trainY <- as.matrix(data.train[,1])
testY <- as.matrix(data.test[,1])
```

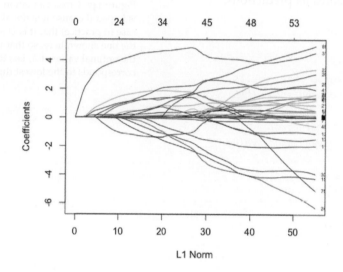

Figure 146: Path trajectory of the fitted regression parameters. The figure should be read from right to left (i.e., λ from small to large). Variables that become zero later are stronger (i.e., since a larger λ is needed to make them become 0). The variables that quickly become zero are weak or insignificant variables.

Step 3 uses the R package `glmnet` [207] to build a LASSO model.

[207] Check out the argument `glmnet` to learn more.

```
# Step 3 -> Use glmnet to conduct LASSO
# install.packages("glmnet")
require(glmnet)
fit = glmnet(trainX,trainY, family=c("gaussian"))

head(fit$beta)
# The fitted sparse regression parameters under
# different lambda values are stored in fit$beta.
```

Step 4 draws the path trajectory of the LASSO models (i.e., as the one shown in Figure 144). The result is shown in Figure 146. It displays the information stored in `fit$beta`. Each curve shows how the estimated regression coefficient of a variable changes according to the value of λ.

```
# Step 4 -> visualization of the path trajectory of
# the fitted sparse regression parameters
plot(fit,label = TRUE)
```

Step 5 uses cross-validation to identify the best λ value for the LASSO model. The result is shown in Figure 147.

```
# Step 5 -> Use cross-validation to decide which lambda to use
cv.fit = cv.glmnet(trainX,trainY)
plot(cv.fit)
# look for the u-shape, and identify the lowest
# point that corresponds to the best model
```

Step 6 views the best model and evaluates its predictions.

```
# Step 6 -> To view the best model and the
# corresponding coefficients
cv.fit$lambda.min
# cv.fit$lambda.min is the best lambda value that results
# in the best model with smallest mean squared error (MSE)
coef(cv.fit, s = "lambda.min")
# This extracts the fitted regression parameters of
# the linear regression model using the given lambda value.

y_hat <- predict(cv.fit, newx = testX, s = "lambda.min")
# This is to predict using the best model
cor(y_hat, data.test$MMSCORE)
mse <- mean((y_hat - data.test$MMSCORE)^2)
# The mean squared error (mse)
mse
```

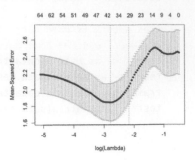

Figure 147: Cross-validation result. It is hoped (because it is not always the case in practice) that it is *U-shaped*, like the one shown here, so that we can spot the optimal value of λ, i.e., the one that corresponds to the lowest dip point.

Results are shown below.

```
## 0.2969686 # cor(y_hat, data.test$MMSCORE)
## 2.453638  # mse
```

Step 7 re-fits the regression model using the variables selected by LASSO. As LASSO put L_1 norm on the regression parameters, it not only penalizes the regression coefficients of the irrelevant variables to be zero, but also penalizes the regression coefficients of the selected variable. Thus, the estimated regression coefficients of a LASSO model tend to be smaller than they are (i.e., this is called *bias* in machine learning terminology).

```
# Step 7 -> Re-fit the regression model with selected variables
# by LASSO
var_idx <- which(coef(cv.fit, s = "lambda.min") != 0)
lm.AD.reduced <- lm(MMSCORE ~ ., data =
                        data.train[,var_idx,drop=FALSE])
summary(lm.AD.reduced)
```

Principal component analysis

Rationale and formulation

A dataset has many variables, but its inherent dimensionality may be smaller than it appears to be. For example, as shown in Figure 148, the 10 variables of the dataset, x_1, x_2, ..., x_{10}, are manifestations of three underlying independent variables, z_1, z_2, and z_3. In other words, a dataset of 10 variables is not necessarily a system of 10 *degrees of freedom*.

The question is how to uncover the "Master of Puppets", i.e., z_1, z_2, and z_3, based on data of the observed variables, x_1, x_2, ..., x_{10}.

Let's look at the scattered data points in Figure 149. If we think of the data points as *stars*, and this is the universe after the *Big Bang*, we can identify two potential forces here: a force that stretches the data points towards one direction (i.e., labeled as *the 1st PC*)[208]; and another force (i.e., labeled as *the 2nd PC*) that drags the data points towards another direction. The forces are independent, so in mathematical terms they follow **orthogonal** directions. And it might be possible that the 2nd PC only represents noise. If that is the case, calling it a force may not be the best way. Sometimes we say each PC represents a *variation source*.

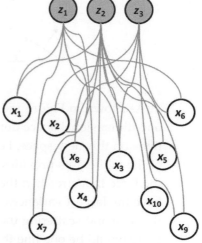

Figure 148: PCA—to uncover the "Master of Puppets" (z_1, z_2, and z_3)

[208] PC stands for the *principal component*.

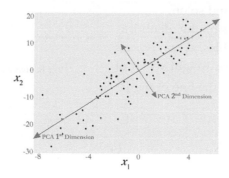

Figure 149: Illustration of the principal components in a dataset with 2 variables; the main variation source is represented by the 1st PC dimension

This interpretation of Figure 149 may seem natural. If so, it is only because it makes a tacit assumption that seems too *natural* to draw our attention: the forces are *represented* as *lines*[209], their mathematical forms are *linear models* that are defined by the existing variables, i.e., the two lines in Figure 149 could be defined by x_1 and x_2. The PCA seeks linear combinations of the original variables to pinpoint the directions towards which the underlying forces push the data points. These directions are called *principal components* (PCs). In other words, the PCA assumes that the relationship between the underlying PCs and the observed variables is linear. And because they are linear, it takes *orthogonality* to separate different forces.

[209] Why is a force a line? It could be a wave, a spiral, or anything other than a line. But the challenge is to write up the mathematical form of an idea—like the example of maximum margin in **Chapter 7**.

Theory and method

The lines in Figure 149 take the form as $w_1x_1 + w_2x_2$,[210] where w_1 and w_2 are free parameters. To estimate w_1 and w_2 for the lines, we need to write an *optimization* formulation with an objective function and a constraints structure that carries out the idea outlined in Figure 149: to identify the two lines.

[210] For simplicity, from now on, we assume that all the variables in the dataset are normalized, i.e., for any variable x_i, its mean is 0 and its variance is 1.

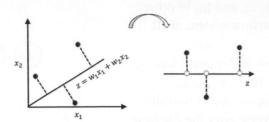

Figure 150: Any line $z = w_1x_1 + w_2x_2$ leads to a new one-dimensional space defined by z

As shown in Figure 150, any line $z = w_1x_1 + w_2x_2$ leads to a new one-dimensional space defined by z. Data points find their projections on this new space, i.e., the white dots on the line. The variance of the white dots provides a quantitative evaluation of the strength of the force that stretched the data points. The PCA seeks the lines that have the largest variances, which are the strongest forces stretching the data and scattering the data points along the PCs. Specifically, as there would be one line that represents the strongest force (a.k.a., as the 1st PC), the second line is called the 2nd PC, and so on.

To generalize the idea of Figure 150, let's focus on the identification of the 1st PC first. Suppose there are p variables, $x_1, x_2, \ldots, x_p$. The *line* for the 1st PC is $w_{(1)}^T x$. $w_{(1)} \in \mathbb{R}^{p \times 1}$ is the weight vector of the 1st PC[211]. The projections of N data points on the line of the 1st PC, i.e., the coordinates of the *white dots*, are

[211] It is also called the **loading** of the PC.

$$z_{1n} = w_{(1)}^T x_n, \text{ for } n = 1, 2, \ldots, N, \qquad (96)$$

where $x_n \in \mathbb{R}^{1 \times p}$ is the nth data point.

As we mentioned, the 1st PC is the line that has the largest variance of z_1. Suppose that the data have been standardized, we have

$$var(z_1) = var\left(w_{(1)}^T x\right) = \frac{1}{N} \sum_{n=1}^{N} \left[w_{(1)}^T x_n\right]^2. \qquad (97)$$

This leads to the following formulation to learn the parameter $w_{(1)}$

$$w_{(1)} = \arg \max_{w_{(1)}^T w_{(1)} = 1} \left\{ \sum_{n=1}^{N} \left[w_{(1)}^T x_n\right]^2 \right\}, \qquad (98)$$

where the constraint $w_{(1)}^T w_{(1)} = 1$ is to normalize the scale of w.[212]

[212] Without which the optimization problem in Eq. 98 is unbounded. This also indicates that the absolute magnitudes of $w_{(1)}$ are often misleading. The relative magnitudes are more useful.

A more succinct form of Eq. 98 is

$$w_{(1)} = \arg \max_{w_{(1)}^T w_{(1)}=1} \left\{ w_{(1)}^T X^T X w_{(1)} \right\}, \qquad (99)$$

where $X \in \mathbb{R}^{N \times p}$ is the data matrix that concatenates the N samples into a matrix, i.e., each sample forms a row in X. Eq. 99 is also known as the **eigenvalue decomposition** problem of the matrix $X^T X$.[213] In this context, $w_{(1)}$ is called the 1st **eigenvector**.

To identify the 2nd PC, we again find a way to *iterate*. The idea is simple: as the 1st PC represents a variance source, and the data X contains an aggregation of multiple variance sources, why not remove the first variance source from X and then create a new dataset that contains the remaining variance sources? Then, the procedure for finding $w_{(1)}$ could be used for finding $w_{(2)}$—with $w_{(1)}$ removed, $w_{(2)}$ is the largest variance source.

This process could be generalized as:

Create $X_{(k)}$ In order to find the kth PC, we could create a dataset by removing the variation sources from the previous $k - 1$ PCs

$$X_{(k)} = X - \sum_{s=1}^{k-1} w_{(s)} w_{(s)}^T. \qquad (100)$$

Solve for $w_{(k)}$ Then, we solve

$$w_{(k)} = \arg \max_{w_{(k)}^T w_{(k)}=1} \left\{ w_{(k)}^T X_{(k)}^T X_{(k)} w_{(k)} \right\}. \qquad (101)$$

We then compute $\lambda_{(k)} = w_{(k)}^T X_{(k)}^T X_{(k)} w_{(k)}$. $\lambda_{(k)}$ is called the **eigenvalue** of the kth PC.

So we create $X_{(k)}$ and solve Eq. 101 in multiple iterations. Many R packages have packed all the iterations into one batch. Usually, we only need to calculate $X^T X$ or S and use it as input of these packages, then obtain all the eigenvalues and eigenvectors.

This iterative algorithm would yield in total p PCs for a dataset with p variables. But usually, not all the PCs are significant. If we apply PCA on the dataset generated by the data-generating mechanism as shown in Figure 148, only the first 3 PCs should be significant, and the other 7 PCs, although they *computationally exist*, statistically do not exist, as they are manifestations of noise.

In practice, we need to decide how many PCs are needed for a dataset. The **scree plot** as shown in Figure 151 is a common tool: it draws the eigenvalues of the PCs, $\lambda_{(1)}, \lambda_{(2)}, \ldots, \lambda_{(p)}$. Then we look for the change point if it exists. We discard the PCs after the change point as they may be statistically insignificant.

[213] $\frac{X^T X}{N-1}$ is called the **sample covariance matrix**, usually denoted as S. S could be used in Eq. 99 to replace $X^T X$.

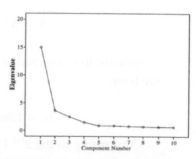

Figure 151: The scree plot shows that only the first 2 PCs are significant

A small data example

The dataset is shown in Table 35. It has 3 variables and 5 data points.

x_1	x_2
−10	6
−4	2
2	1
8	0
14	−4

Table 35: A dataset example for PCA

First, we normalize (or, standardize) the variables[214]. I.e., for x_1, we compute its mean and standard derivation first, which are 2 and 9.48,[215] respectively. Then, we distract each measurement of x_1 from its mean and further divide it by its standard derivation. For example, for the first measurement of x_1, −10, it is converted as

$$\frac{-10 - 2}{9.48} = -1.26.$$

The second measurement, −4, is converted as

$$\frac{-4 - 2}{9.48} = -0.63.$$

And so on.

Similarly, for x_2, we compute its mean and standard derivation, which are 1 and 3.61, respectively. The standardized dataset is shown in Table 36.

x_1	x_2
−1.26	1.39
−0.63	0.28
0	0
0.63	−0.28
1.26	−1.39

[214] Recall that we assumed that all the variables are normalized when we derived the PCA algorithm

[215] In this example, numbers are rounded to 2 decimal places.

We calculate S as

$$S = X^T X / 4 = \begin{bmatrix} 1 & -0.96 \\ -0.96 & 1 \end{bmatrix}.$$

Solving this eigenvalue decomposition problem[216], for the 1st PC, we have

$$\lambda_1 = 1.96 \text{ and } w_{(1)} = [-0.71, 0.71].$$

Continuing to the 2nd PC, we have

$$\lambda_2 = 0.04 \text{ and } w_{(2)} = [-0.71, -0.71].$$

[216] E.g., using `eigen()` in R.

Table 36: Standardized dataset of Table 35

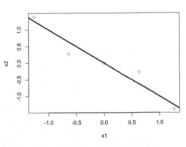

Figure 152: Gray dots are data points (standardized); the black line is the 1st PC

We can calculate the cumulative contributions of the 2 PCs

$$\text{For the } 1^{st} \text{ PC: } 1.96/(1.96 + 0.04) = 0.98.$$

$$\text{For the } 2^{nd} \text{ PC: } 0.04/(1.96 + 0.04) = 0.02.$$

The 2nd PC is statistically insignificant.

We visualize the 1st PC in Figure 152 (compare it with Figure 149). The R code to generate Figure 152 is shown below.

```
x1 <- c(-10, -4, 2, 8, 14)
x2 <- c(6, 2, 1, 0, -4)
x <- cbind(x1,x2)
x.scale <- scale(x) #standardize the data
eigen.x <- eigen(cor(x))
plot(x.scale, col = "gray", lwd = 2)
abline(0,eigen.x$vectors[2,1]/eigen.x$vectors[1,1],
       lwd = 3, col = "black")
```

The coordinates of the *white dots* (a.k.a., the projections of the data points on the PCs, as shown in Figure 150) can be obtained by using Eq. 96. Results are shown in Table 37. This is an example of *data transformation*.

z_1	z_2
1.88	−0.09
0.64	0.25
0	0
−0.64	−0.25
−1.88	0.09

Table 37: The coordinates of the *white dots*, i.e., a.k.a., the projections of the data points on the PCs

Data transformation is often a data preprocessing step before the use of other methods. For example, in clustering, sometimes we could not discover any clustering structure on the dataset of original variables, but we may discover clusters on the transformed dataset. In a regression model, as we have mentioned the issue of multicollinearity[217], the **Principal Component Regression (PCR)** method uses the PCA first to convert the original x variables into the z variables and then applies the linear regression model on the transformed variables. This is because the z variables are PCs and they are orthogonal with each other, without issue of multicollinearity.

R Lab

The 6-Step R Pipeline   **Step 1** and **Step 2** get dataset into R and organize the dataset in the required format [218].

[217] I.e., in **Chapter 6** and **Chapter 2**.

[218] It is not necessary to split the dataset into training and testing datasets before the use of PCA, *if* the purpose of the analysis is *exploratory data analysis*. But if the purpose of using PCA is for *dimension reduction* or *feature extraction*, which is an intermediate step before building a prediction model, then we should split the dataset into training and testing datasets, and apply PCA only on the training dataset to learn the loadings of the significant PCs. The R lab shows an example of this process.

```
# Step 1 -> Read data into R
#### Read data from a CSV file
#### Example: Alzheimer's Disease

# RCurl is the R package to read csv file using a link
library(RCurl)
url <- paste0("https://raw.githubusercontent.com",
              "/analyticsbook/book/main/data/AD_hd.csv")
AD <- read.csv(text=getURL(url))
# str(AD)
```

```
# Step 2 -> Data preprocessing
# Create your X matrix (predictors) and Y vector
# (outcome variable)
X <- AD[,-c(1:16)]
Y <- AD$MMSCORE

# Then, we integrate everything into a data frame
data <- data.frame(Y,X)
names(data)[1] = c("MMSCORE")

# Create a training data
train.ix <- sample(nrow(data),floor( nrow(data)) * 4 / 5 )
data.train <- data[train.ix,]
# Create a testing data
data.test <- data[-train.ix,]

trainX <- as.matrix(data.train[,-1])
testX <- as.matrix(data.test[,-1])
trainY <- as.matrix(data.train[,1])
testY <- as.matrix(data.test[,1])
```

Step 3 implements the PCA analysis using the `FactoMineR` package.

```
# Step 3 -> Implement principal component analysis
# install.packages("factoextra")
require(FactoMineR)
# Conduct the PCA analysis
pca.AD <- PCA(trainX,  graph = FALSE,ncp=10)
# names(pca.AD) will give you the list of variable names in the
# object pca.AD created by PCA(). For instance, pca.AD$eig records
# the eigenvalues of all the PCs, also the transformed value into
# cumulative percentage of variance. pca.AD$var stores the
# loadings of the variables in each of the PCs.
```

Step 4 ranks the PCs based on their eigenvalues and identifies the significant ones.

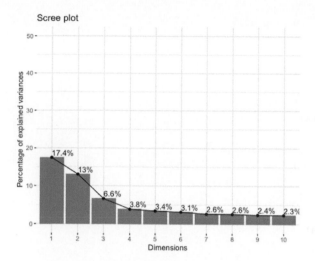

Figure 153: Scree plot of the PCA analysis on the AD dataset

```
# Step 4 -> Examine the contributions of the PCs in explaining
# the variation in data.
require(factoextra )
# to use the following functions such as get_pca_var()
# and fviz_contrib()
fviz_screeplot(pca.AD, addlabels = TRUE, ylim = c(0, 50))
```

The result is shown in Figure 153. The 1st PC explains away 17.4% of the total variation, and the 2nd PC explains away 13% of the total variation. There is a change point at the 3rd PC, indicating that the following PCs may be insignificant.

Step 5 looks into the details of the learned PCA model, e.g., the *loadings* of the PCs. It leads to Figures 154 and 155 which visualize the contributions of the variables to the 1st and 2nd PC, respectively.

```
# Step 5 -> Examine the loadings of the PCs.
var <- get_pca_var(pca.AD) # to get the loadings of the PCs
head(var$contrib) # to show the first 10 PCs

# Visualize the contributions of top variables to
# PC1 using a bar plot
fviz_contrib(pca.AD, choice = "var", axes = 1, top = 20)
# Visualize the contributions of top variables to PC2 using
# a bar plot
fviz_contrib(pca.AD, choice = "var", axes = 2, top = 20)
```

Step 6 implements linear regression model using the transformed data.

```
# Step 6 -> use the transformed data fit a line regression model

# Data pre-processing
```

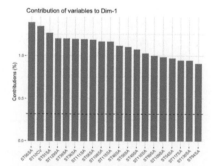

Figure 154: Loading of the 1st PC, i.e., coefficients are ranked in terms of their absolute magnitude and only the top 20 are shown

```
# Transformation of the X matrix of the training data
trainX <- pca.AD$ind$coord
trainX <- data.frame(trainX)
names(trainX) <- c("PC1","PC2","PC3","PC4","PC5","PC6","PC7",
                   "PC8","PC9","PC10")
# Transformation of the X matrix of the testing data
testX <- predict(pca.AD , newdata = testX)
testX <- data.frame(testX$coord)
names(testX) <- c("PC1","PC2","PC3","PC4","PC5","PC6",
                  "PC7","PC8","PC9","PC10")


tempData <- data.frame(trainY,trainX)
names(tempData)[1] <- c("MMSCORE")
lm.AD <- lm(MMSCORE ~ ., data = tempData)
summary(lm.AD)


y_hat <- predict(lm.AD, testX)
cor(y_hat, testY)
mse <- mean((y_hat - testY)^2) # The mean squared error (mse)
mse
```

The result is shown below.

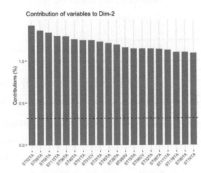

Figure 155: Loading of the 2nd PC, i.e., coefficients are ranked in terms of their absolute magnitude and only the top 20 are shown

```
##
## Call:
## lm(formula = AGE ~ ., data = tempData)
##
## Residuals:
##     Min      1Q  Median      3Q     Max
## -17.3377 -2.5627  0.0518  2.6820 11.1772
##
## Coefficients:
##             Estimate Std. Error t value Pr(>|t|)
## (Intercept) 73.68767    0.59939 122.938  < 2e-16 ***
## PC1          0.04011    0.08275   0.485 0.629580
## PC2         -0.31556    0.09490  -3.325 0.001488 **
## PC3          0.50022    0.13510   3.702 0.000456 ***
## PC4          0.14812    0.17462   0.848 0.399578
## PC5          0.47954    0.19404   2.471 0.016219 *
## PC6         -0.29760    0.20134  -1.478 0.144444
## PC7          0.10160    0.21388   0.475 0.636440
## PC8         -0.25015    0.22527  -1.110 0.271100
## PC9         -0.02837    0.22932  -0.124 0.901949
## PC10         0.16326    0.23282   0.701 0.485794
## ---
## Signif. codes:  0 '***' 0.001 '**' 0.01 '*' 0.05 '.' 0.1 ' ' 1
##
## Residual standard error: 5.121 on 62 degrees of freedom
## Multiple R-squared:  0.3672, Adjusted R-squared:  0.2651
## F-statistic: 3.598 on 10 and 62 DF,  p-value: 0.0008235
```

It is not uncommon to see that the 1st PC is insignificant in a prediction model. The 1st PC is the largest *force* or *variation source* in X by definition, but not necessarily the one that correlates with any outcome variable y with the strongest correlation.

On the other hand, the *R-squared* of this model is 0.3672, and the *p-value* is 0.0008235. Overall, the data transformation by PCA yielded an effective linear regression model.

Beyond the 6-Step R Pipeline   PCA is a popular tool for *EDA*. For example, we can visualize the distribution of the data points in the new space spanned by a few selected PCs[219]. We use the following R script to draw a visualization figure.

[219] It may reveal some structures of the dataset. For example, for a classification problem, it is hoped that in the space spanned by the selected PCs the data points from different classes would cluster around different centers.

```
# Projection of data points in the new space defined by
# the first two PCs
fviz_pca_ind(pca.AD, label="none",
             habillage=as.factor(AD[train.ix,]$DX_bl),
             addEllipses=TRUE, ellipse.level=0.95)
```

The result is shown in Figure 156. Two clusters are identified, which overlap significantly. One group is the *LMCI* (i.e., mild cognitive impairment) and the other one is *NC* (i.e., normal aging). The result is consistent with the fact that the clinical difference between the two groups is not as significant as *NC* versus *Diseased*.

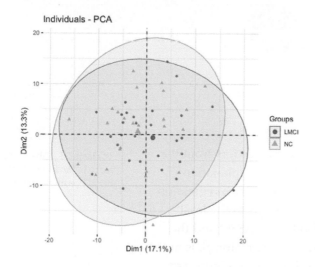

Figure 156: Scatterplot of the subjects in the space defined by the 1st and 2nd PCs

Remarks

Why LASSO uses the L_1 norm

LASSO is often compared with another model, the **Ridge regression** that was developed about 30 years before LASSO[220].

[220] Hoerl, A.E. and Kennard, R.W. *Ridge regression: biased estimation for nonorthogonal problems*, Technometrics, Volume 12, Issue 1, Pages 55-67, 1970.

The formulation of Ridge regression is

$$\hat{\beta} = \arg\min_{\beta} \left\{ \underbrace{(y - X\beta)^T (y - X\beta)}_{\text{Least squares}} + \underbrace{\lambda \|\beta\|_2^2}_{L_2 \text{ norm penalty}} \right\} \quad (102)$$

where $\|\beta\|_2^2 = \sum_{i=1}^{p} |\beta_i|^2$.

Ridge regression seems to bear the same spirit of LASSO—they both penalize the magnitudes of the regression parameters. However, it has been noticed that in the Ridge regression model the estimated regression parameters are less likely to be 0. Even with a very large λ, many elements in β may be close to zero (i.e., with a tiny numerical magnitude), but not zero[221]. This may not be entirely a surprise, as the Ridge regression is often used as a *stabilization* strategy to handle the *multicollinearity* issue or other issues that result in numerical instability of parameter estimation in linear regression, while LASSO is mainly used as a *variable selection* strategy.

[221] If they are not zero, these variables can still generate impact on the estimation of other regression parameters.

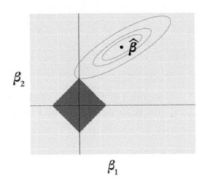 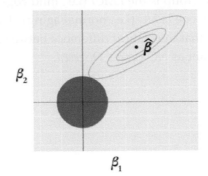

Figure 157: Why LASSO (left) generates sparse estimates, while Ridge regression (right) does not

To reveal why the L_1 norm in LASSO regression differs from the L_2 norm used in the Ridge regression, we adopt an explanation[222] as shown in Figure 157. There are 2 predictors, thus, two regression coefficients β_1 and β_2. The contour plot corresponds to the least squares loss function which is shared by both the LASSO and the Ridge regression models. And the least squares estimator, $\hat{\beta}$, is in the center of the contour plots. The shadowed rhombus in Figure 157 (left) corresponds to the L_1 norm, and the shadowed circle in Figure 157 (right) corresponds to the L_2 norm. For either model, the optimal solution happens at the *contact point* of the two shapes.

Figure 157 shows that the contact point of the elliptic contour plot with the shadowed rhombus is likely to be one of the *sharp* corner points. A feature of these corner points is that some variables are zero, e.g., in Figure 157 (left), the point of contact implies that $\beta_1 = 0$.

[222] Hastie, T., Tibshirani, R. and Friedman, J. *The Elements of Statistical Learning*, 2nd edition. Springer, 2009.

As a comparison, in Ridge regression, the shadowed circle has no such sharp corner points. Given the infinite number of potential contact points of the elliptic contour plot with the shadowed circle, it is expected that the Ridge regression will not result in sparse solutions with exact zeros in the estimated regression coefficients.

Following this idea[223], the L_1 norm is extended to the L_q norm, where $q \leq 1$. For any $q \leq 1$, we could generate sharp corner points to enable sparse solutions. The advantage of using $q < 1$ is to reduce bias in the model[224]. The cost of using $q < 1$ is that it will result in *non-convex* penalty terms, creating a more challenging optimization problem than LASSO. Considerable amounts of efforts have been devoted to two main directions: development of new norms, and development of new algorithms (i.e., which are usually iterative procedures with closed-form solution in each iteration, like the Shooting algorithm). Interested readers can read more of these works[225].

The myth of PCA

While PCA has been widely used, it is often criticized as a *black box* model that lacks *interpretability*. It depends on the circumstances where the PCA is used. Sometimes, it is not easy to connect the identified principal components with physical entities. The applications of PCA in many areas have formed a convention, or a myth—some statisticians may say—such that formulistic rubrics have been invented to convert their data into patterns, then further convert these patterns into formulated sentences such as, "the variables that have larger magnitudes in the first 3 PCs correspond to the brain regions in the hippocampus area, indicating that these brain regions manifest significant functional connectivity to deliver the verbal function", or "we have identified 5 significant PCs, and the genes that show dominant magnitudes in the loading of the 1st PC are all related to T-cell production and immune functions ... each of the PC indicates a biological pathway that consists of these constitutional genes working together to produce specific types of proteins". Then hear what had been said by financial analysts: "using PCA on 100 stocks[226], we found that the 1st PC consists of 10 stocks as their weights in the loading are significantly larger than the other stocks. This may indicate that there is strong correlation between these 10 stocks ... consider this fact when you come up with your investment strategy..."

Having said that, sometimes there is magic in PCA.

Consider another small data example that is shown in Table 38. It has 3 variables and 8 data points.

First, we normalize the variables, i.e., for x_1, we compute its mean and standard derivation first, which are 2.375 and 3.159, respectively.

[223] I.e., to create sharp contact points between the elliptical contour with the shape representing the norm.

[224] I.e., the L_1 norm not only penalizes the regression coefficients of the irrelevant variables to be zero, it also penalizes the regression coefficients of the relevant variables. This is a *bias* in the model.

[225] A good place to start with: `https://github.com/jiayuzhou/SLEP` and its manual (in PDF).

[226] Each stock is a variable.

x_1	x_2	x_3
−1	0	1
3	3	1
3	5	1
−3	−2	1
3	4	1
5	6	1
7	6	1
2	2	0

Table 38: A dataset example for PCA

Then, we distract each measurement of x_1 from its mean and further divide it by its standard derivation. For example, for the first measurement of x_1, −1, it is converted as

$$\frac{-1 - 2.375}{3.159} = -1.07.$$

The second measurement, 3, is converted as

$$\frac{3 - 2.375}{3.159} = 0.20.$$

And so on.

Similarly, for x_2, we compute its mean and standard derivation, which are 3 and 2.88, respectively. For x_3, we compute its mean and standard derivation, which are 0.88 and 0.35, respectively .... Then the standardized dataset is shown in Table 39.[227]

[227] In this example, numbers are rounded to 2 decimal places.

Table 39: Standardized dataset of Table 38

x_1	x_2	x_3
−1.07	−1.04	0.35
0.2	0.00	0.35
0.2	0.69	0.35
−1.70	−1.73	0.35
0.20	0.35	0.35
0.83	1.04	0.35
1.46	1.04	0.35
−0.11	−0.35	−2.48

We calculate the sample covariance matrix[228] S as

$$S = \frac{X^T X}{N - 1} = \begin{bmatrix} 1 & 0.96 & 0.05 \\ 0.96 & 1 & 0.14 \\ 0.05 & 0.14 & 1 \end{bmatrix}.$$

By solving the eigenvalue decomposition problem of the matrix S, we obtain the PCs and their loadings.

For the 1st PC, we get

$$\lambda_1 = 1.98 \text{ and } w_{(1)} = [-0.69, -0.70, -0.14].$$

[228] From S we see that the correlation between x_1 and x_2 is quite large, while the correlation between them with x_3 is very small. Figure 158 visualizes this relationship of the three variables.

Figure 158: Visualization of the relationship between the three variables

For the 2nd PC, we get

$$\lambda_2 = 0.98 \text{ and } w_{(2)} = [0.14, 0.05, -0.99].$$

For the 3rd PC, we get

$$\lambda_3 = 0.04 \text{ and } w_{(3)} = [0.70, -0.71, 0.07].$$

We can calculate the cumulative contributions of the three PCs

For the 1st PC: $1.98/(1.98 + 0.98 + 0.04) = 0.66$.

For the 2nd PC: $0.98/(1.98 + 0.98 + 0.04) = 0.33$.

For the 3rd PC: $0.04/(1.98 + 0.98 + 0.04) = 0.01$.

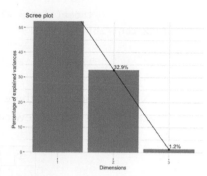

Figure 159: Scree plot of the PCA analysis on data in Table 39

The 3rd PC is statistically insignificant. The scree plot is shown in Figure 159.

We look into the details of the learned PCA model, e.g., the *loadings* of the PCs. It leads to Figure 160.

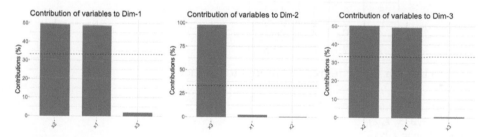

Figure 160: Loadings of the 1st PC (left), 2nd PC (middle), and 3rd PC (right)

Figure 160 shows that the 1st PC is mainly defined by x_1 and x_2, the 2nd PC is mainly defined by x_3, and the 3rd PC, despite its small proportion of importance, mainly consists of x_1 and x_2 as well[229].

The R code for generating Figure 160 is shown below.

[229] Readers may compare Figure 160 with Figure 158—Is this a coincidence?

```
# PCA example
x1 <- c(-1,3,3,-3,3,5,7,2)
x2 <- c(0,3,5,-2,4,6,6,2)
x3 <- c(1,1,1,1,1,1,1,0)
X <- cbind(x1,x2,x3)
require(FactoMineR)
require(factoextra)
t <- PCA(X)
t$eig
t$var$coord

# Draw the screeplot
```

```
fviz_screeplot(t, addlabels = TRUE)

# Draw the variable loadings plot
fviz_contrib(t, choice = "var", axes = 1, top = 20,
             sort.val = "none") +
        theme(text = element_text(size = 20))
fviz_contrib(t, choice = "var", axes = 2, top = 20,
             sort.val = "none") +
        theme(text = element_text(size = 20))
fviz_contrib(t, choice = "var", axes = 3, top = 20,
             sort.val = "none") +
        theme(text = element_text(size = 20))
```

Now, suppose that there is an outcome variable y. Data in Table 38 is augmented with a new column, as shown in Table 40.

x_1	x_2	x_3	y
−1	0	1	1.33
3	3	1	0.70
3	5	1	2.99
−3	−2	1	−1.78
3	4	1	0.07
5	6	1	4.62
7	6	1	3.87
2	2	0	0.58

Table 40: Table 38 is augmented with an outcome variable

The goal is to build a linear regression model to predict y.

```
# Build a linear regression model
x1 <- c(-1,3,3,-3,3,5,7,2)
x2 <- c(0,3,5,-2,4,6,6,2)
x3 <- c(1,1,1,1,1,1,1,0)
X <- cbind(x1,x2,x3)
y <- c(1.33,0.7,2.99,-1.78,0.07,4.62,3.87,0.58)
data <- data.frame(cbind(y,X))
lm.fit <- lm(y~., data = data)
summary(lm.fit)
```

The result is shown below.

```
## Call:
## lm(formula = y ~ ., data = data)
## Coefficients:
##             Estimate Std. Error t value Pr(>|t|)
## (Intercept) -0.64698    1.60218  -0.404    0.707
## x1          -0.03686    0.67900  -0.054    0.959
## x2           0.65035    0.75186   0.865    0.436
## x3           0.37826    1.75338   0.216    0.840
##
```

```
## Residual standard error: 1.546 on 4 degrees of freedom
## Multiple R-squared:  0.6999, Adjusted R-squared:  0.4749
## F-statistic:  3.11 on 3 and 4 DF,  p-value: 0.1508
```

The *R-squared* is 0.6999, but the three variables are not significant. This is unusual. Recall that x_1 and x_2 are highly correlated—there is an issue of *multicollinearity* in this dataset.

Try a linear regression model with x_1 and x_3.

```
# Build a linear regression model
lm.fit <- lm(y~ x1 + x2, data = data)
summary(lm.fit)
```

The result is shown below.

```
## Call:
## lm(formula = y ~ ., data = data)
## Coefficients:
##             Estimate Std. Error t value Pr(>|t|)
## (Intercept)  -0.4764     1.5494  -0.307   0.7709
## x1            0.5282     0.1805   2.927   0.0328 *
## x3            0.8793     1.6127   0.545   0.6090
##
## Residual standard error: 1.507 on 5 degrees of freedom
## Multiple R-squared:  0.6438, Adjusted R-squared:  0.5013
## F-statistic: 4.519 on 2 and 5 DF,  p-value: 0.07572
```

Now x_1 is significant. Without fitting another model, we know that x_2 has to be significant as well, i.e., as shown in Figure 158, x_1 and x_2 are two highly correlated variables that are just like one variable. But because of the multicollinearity, when both are included in the model, neither turns out to be significant.

To overcome the multicollinearity, we have mentioned that the Principal Component Regression (PCR) method is a good approach. First, we calculate the transformed data (i.e., the projections of the data points on the PCs, as shown in Figure 150) using Eq. 96. Results are shown in Table 41. An important characteristic of the new variables, PC_1, PC_2, and PC_3, is that they are orthogonal to each other, and thus, their correlations are 0.

Then we can build a linear regression model of y using the three new predictors, PC_1, PC_2, and PC_3. The result is shown below.

```
## Call:
## lm(formula = y ~ PC1 + PC2 + PC3, data = data)
## Coefficients:
##             Estimate Std. Error t value Pr(>|t|)
## (Intercept)  1.54750    0.54668   2.831   0.0473 *
## PC1         -1.25447    0.41571  -3.018   0.0393 *
```

PC$_1$	PC$_2$	PC$_3$
1.43	−0.55	0.01
−0.18	−0.32	0.16
−0.67	−0.29	−0.33
2.36	−0.68	0.06
−0.43	−0.30	−0.08
−1.36	−0.18	−0.13
−1.80	−0.09	0.31
0.66	2.41	−0.01

Table 41: The coordinates of the *white dots*, i.e., aka, the projections of the data points on the PCs

```
## PC2         -0.06022    0.58848  -0.102   0.9234
## PC3         -1.39950    3.02508  -0.463   0.6677
##
## Residual standard error: 1.546 on 4 degrees of freedom
## Multiple R-squared:  0.6999,  Adjusted R-squared:  0.4749
## F-statistic:  3.11 on 3 and 4 DF,   p-value: 0.1508
```

PC$_1$ is significant. Since PC$_1$ is mainly defined by x_1 and x_2, this is consistent with all the analysis done so far, and a structure of the relationships between the variables is revealed in Figure 161.

Exercises

1. Figure 162 shows the path trajectory generated by applying `glmnet()` on a dataset with 10 predictors. Which two variables are the top two significant variables (note the index of the variables is shown in the right end of the figure)?

2. Consider the dataset shown in Table 42. Set $\lambda = 1$ and initial values for $\beta_1 = 0$, and $\beta_2 = 1$. Implement the Shooting algorithm by manual operation. Do one iteration. Report β_1 and β_2.

x_1	x_2	y
−0.15	−0.48	0.46
−0.72	−0.54	−0.37
1.36	−0.91	−0.27
0.61	1.59	1.35
−1.11	0.34	−0.11

3. Follow up on the dataset in Q2. Use the R pipeline for LASSO on this data. Compare the result from R and the result by your manual calculation.

4. Conduct a principal component analysis for the dataset shown in Table 43. Show details of the process.

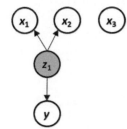

Figure 161: Visualization of the relationship between all the variables

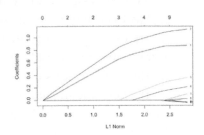

Figure 162: The path trajectory of a LASSO model

Table 42: Dataset for Q2

x_1	x_2	x_3	x_4
1	1.8	2.08	−0.28
2	3.6	−0.78	0.79
1	2.2	−0.08	−0.52
2	4.3	0.38	−0.47
1	2.1	0.71	1.03
2	3.6	1.29	0.67
1	2.2	0.57	0.15
2	4.0	1.12	1.18

Table 43: Dataset for Q4

(a) Standardize the dataset (i.e., by making the means of the variables to be zero, and the standard derivations of the variables to be 1). (b) Calculate the sample covariance matrix (i.e., $S = (X^T X)/(N − 1)$). (c) Conduct eigenvalue decomposition on the sample covariance matrix, obtain the four eigenvectors and their eigenvalues. (d) Report the percentage of variances that could be explained by the four PCs, respectively. Draw the screeplot. How many PCs are sufficient to represent the dataset? In other words, which PCs are significant? (e) Interpret the PCs you have selected, i.e., which variables define which PCs? (f) Convert the original data into the space spanned by the four PCs, by filling in Table 44.

PC_1	PC_2	PC_3	PC_4

Table 44: Dataset for Q4

5. Follow up on the dataset in Q2 from Chapter 7. (a) Conduct the PCA analysis on the three predictors to identify the three principal components and their contributions on explaining the variance in data; and (b) use the R pipeline for PCA to do the PCA analysis and compare with your manual calculation.

6. Suppose that we have an outcome variable that could be augmented into the dataset in Q4, as shown in Table 45. Apply the shooting algorithm for LASSO on this dataset to identify important variables. Use the following initial values for the parameters, $\lambda = 1, \beta_1 = 0, \beta_2 = 1, \beta_3 = 1, \beta_4 = 1$, and just do one iteration of the shooting algorithm. Show details of manual calculation.

x_1	x_2	x_3	x_4	y
1	1.8	2.08	−0.28	1.2
2	3.6	−0.78	0.79	2.1
1	2.2	−0.08	−0.52	0.8
2	4.3	0.38	−0.47	1.5
1	2.1	0.71	1.03	0.8
2	3.6	1.29	0.67	1.6
1	2.2	0.57	0.15	1.2
2	4.0	1.12	1.18	1.6

Table 45: Dataset for Q6

7. After extraction of the four PCs from Q4, use `lm()` in R to build a linear regression model with the outcome variable (as shown in Table 45) and the four PCs as the predictors. (a) Report the summary of your linear regression model with the four PCs; and (b) which PCs significantly affect the outcome variable?

8. Revisit Q1 in **Chapter 3**. Derive the shooting algorithm for weighted least squares regression with L_1 norm penalty.

9. Design a simulated experiment to evaluate the effectiveness of the `glmet()` in the R package `glmnet`. (a) For instance, you can simulate 20 samples from a linear regression model with 10 variables, where only 2 out of the 10 variables are truly significant, e.g., the true model is

$$y = \beta_1 x_1 + \beta_2 x_2 + \epsilon,$$

where $\beta_1 = 1$, $\beta_2 = 1$, and

$$\epsilon \sim N(0,1).$$

You can simulate x_1 and x_2 using the standard normal distribution $N(0,1)$. For the other 8 variables, x_3 to x_{10}, you can simulate each from $N(0,1)$. In data analysis, we will use all 10 variables as predictors, since we won't know the true model. (b) Run `lm()` on the simulated data and comment on the results. (c) Run `glmnet()` on the simulated data, and check the path trajectory plot to see if the true significant variables could be detected. (d) Use the cross-validation process integrated into the `glmnet` package to see if the true significant variables could be detected. (e) Use `rpart()` to build a decision tree and extract the variable importance score to see if the true significant variables could be detected. (f) Use `randomforest()` to build a random forest model and extract the variable importance score, to see if the true significant variables could be detected.

Chapter 9: Pragmatism
Experience & Experimental

Overview

Chapter 9 is about *pragmatism*. Pragmatism is an interesting word because, sometimes, being pragmatic means the opposite of *"-ism"*. Read what was said about pragmatism: *"Consider the practical effects of the objects of your conception. Then, your conception of those effects is the whole of your conception of the object".*[230] In a sense, this resonates with*"why by their fruits you shall know them."*[231] To analyze a dataset, we shall be aware of a tendency that we often apply concepts before we see the dataset. EDA has provided a practical way to help us see our preconceptions. In this chapter, two more methods are presented, including the **kernel regression model** that generalizes the idea of linear regression, and the **conditional variance regression model** that creates a convolution by using regression twice.

[230] Peirce, C. S., *How to Make Our Ideas Clear*, Popular Science Monthly, v. 12, 286–302, 1878.

[231] Matthew 7:20.

Kernel regression model

Rationale and formulation

Simple models, similar to the linear regression model, are like *parents who tell white lies*. A model is simple, not only in the sense that it looks simple, but also because it builds on assumptions that simplify reality. Among all the simple models, the linear regression model is particularly good at disguising its simplicity—it seems so natural that we often forget that its simplicity is its assumption. Simple in its cosmology, not necessary in its terminology—that is what the phrase *simple model* means[232].

"*Simple, but not simpler*" said Albert Einstein.

One such assumption the linear regression model has made, obviously, is *linearity*. It would be OK in practice, as Figure 6 in **Chapter 2** assures us: it is not perfect but it is a good approximation.

Now let's look at Figure 163 (left). The *true model*, represented by

[232] In this sense, a model, regardless of how sophisticated its mathematical representation is, is a simple model if its assumptions simplify reality to such an extent that demands our leap of faith.

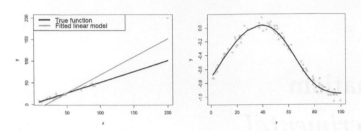

Figure 163: (Left) A single outlier (a local pattern) could impact the regression model as a whole; (right) a *localized* regression model (i.e., the curvature adapts to the locality instead of following a prescribed global form such as a straight line)

the black line, is truly a line. But the fitted model, the orange line, deviates from the black line. In other words, even if the linearity assumption is correct, the consequence is not what we hope for.

The troublemaker appears to be the outlier located on the upper right corner of the figure. As discernible data scientists, we should be aware that the *model* is general, while the *dataset* at hand is particular. It is OK to say the outlier is the troublemaker, but note that this outlier is accidental. The real troublemaker is what *enables* the possibility of outlier to be a troublemaker. The real troublemaker lies deeper.

A common theme of the methods in this book is to establish certainty in a world of uncertainty. The linearity assumption is an assumption, since real data rarely give you a perfect straight line. The way it deals with uncertainty is to use the least squares principle for model estimation. It aims to look for a line that could pierce through *all* the data points. This makes each data point have a *global* impact: mentally move any data point in Figure 163 (left) up and down and imagine how the fitted orange line would move up and down accordingly. In other words, as a data point in any location could change the line dramatically, the linear regression model, together with its least squares estimation method, has imposed an even stronger assumption than merely linearity: it assumes that knowledge learned from one location would be universally useful to all other locations. This implicit assumption[233] could be irrational in some applications, where the data points collected in one location may only tell information about that local area, not easily generalizable to the whole space. Thus, when the global models fail, we need *local models*[234] to fit the data, as shown in Figure 163 (right).

[233] I.e., models that have made this assumption are often termed as *global models*.

[234] A *local model* more relies on the data points in a neighborhood to build up the part of the curve that comes through that particular neighborhood.

Theory and method

Suppose there are N data points, denoted as, (x_n, y_n) for $n = 1, 2, \ldots, N$. To predict on a point x^*, a *local model* assumes the following structure

$$y^* = \sum_{n=1}^{N} y_n w(x_n, x^*). \tag{103}$$

Here, $w(x_n, x^*)$ is the **weight function** that characterizes the *similarity* between x^* and the training data points, x_n, for $n = 1, 2, \ldots, N$. The idea is to predict on a data point based on the data points that are nearby. Methods differ from each other in terms of how they define $w(x_n, x^*)$.

Roughly speaking, there are two main methods. One is the **K-nearest neighbor (KNN)** smoother, and another is the **kernel** smoother.

The KNN smoother   The KNN smoother defines $w(x_n, x^*)$ as

$$w(x_n, x^*) = \begin{cases} \frac{1}{k}, & \text{if } x_n \text{ is one of the } k \text{ nearest neighbors of } x^*; \\ 0, & \text{if } x_n \text{ is NOT among the } k \text{ nearest neighbors of } x^*. \end{cases}$$

Here, to define the *nearest neighbors* of a data point, a *distance function* is needed. Examples include the *Euclidean*[235], *Mahalanobis*, and *Cosine* distance functions. What distance function to use depends on the characteristics of the data. Model selection methods such as the cross-validation can be used to select the best distance function for a dataset.

[235] E.g., $d(x_n, x_m) = \sqrt{\sum_{i=1}^{p} (x_{ni} - x_{mi})^2}$.

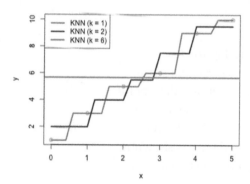

Figure 164: Three KNN smoother models ($k = 1$, $k = 2$, and $k = 6$)

Consider a data example as shown in Table 46. A visualization of the data points is shown in Figure 164, i.e., the gray data points.

ID	x	y
1	0	1
2	1	3
3	2	5
4	3	6
5	4	9
6	5	10

Table 46: Example of a dataset with 6 data points

Let's build 3 KNN smoother models (e.g., $k = 1$, $k = 2$, and $k = 6$) and use the Euclidean distance function to identify the *nearest neighbors* of a data point. Results are presented in Tables 47, 48, and 49, respectively. Note that, in this dataset, as there are in total 6 data points, the KNN model with $k = 6$ is the same as the trivial model that uses the average of y as predictions for all data points.

x^*	KNN	y^*
0.4	x_1	y_1
1.6	x_3	y_3
3.2	x_4	y_4
4.8	x_6	y_6

Table 47: Predictions by a KNN smoother model with $k = 1$ on some locations of x^*

x^*	KNN	y^*
0.4	x_1, x_2	$(y_1 + y_2)/2$
1.6	x_2, x_3	$(y_2 + y_3)/2$
3.2	x_4, x_5	$(y_4 + y_5)/2$
4.8	x_5, x_6	$(y_5 + y_6)/2$

Table 48: Predictions by a KNN smoother model with $k = 2$ on some locations of x^*

x^*	KNN	y^*
0.4	x_1-x_6	$\sum_{n=1}^{6} y_n/6$
1.6	x_1-x_6	$\sum_{n=1}^{6} y_n/6$
3.2	x_1-x_6	$\sum_{n=1}^{6} y_n/6$
4.8	x_1-x_6	$\sum_{n=1}^{6} y_n/6$

Table 49: Predictions by a KNN smoother model with $k = 6$ on some locations of x^*

The 3 KNN smoother models are also shown in Figure 164.

A distinct feature of the **KNN smoother** is the *discrete* manner to define the similarity between data points, which is, for any data point x^*, the data point x_n is either a neighbor or not. The KNN smoother only uses the k nearest neighbors of x^* to predict y^*. This discrete manner of the KNN smoother results in the serrated curves shown in Figure 164. This is obviously artificial, pointing out a systematic *bias* imposed by the KNN smoother model.

The kernel smoother   To remove this bias, the **kernel smoother** creates *continuity* in the similarity between data points. A kernel smoother defines $w(x_n, x^*)$ in the following manner

$$w\left(x_n, x^*\right) = \frac{K\left(x_n, x^*\right)}{\sum_{n=1}^{N} K\left(x_n, x^*\right)}.$$

Here, $K\left(x_n, x^*\right)$ is a *kernel function* as we have discussed in **Chapter 7**. There have been many kernel functions developed, for example, as shown in Table 50.

Many kernel functions are smooth functions. To understand a kernel function, using R to draw it is a good approach. For example, the following R code draws a few instances of the *Gaussian radial basis*

Kernel Function	Mathematical Form	Parameters
Line	$K(x_i, x_j) = x_i^T x_j$	*null*
Polynomial	$K(x_i, x_j) = \left(x_i^T x_j + 1\right)^q$	q
Gaussian radial basis	$K(x_i, x_j) = e^{-\gamma \|x_i - x_j\|^2}$	$\gamma \geq 0$
Laplace radial basis	$K(x_i, x_j) = e^{-\gamma \|x_i - x_j\|}$	$\gamma \geq 0$
Hyperbolic tangent	$K(x_i, x_j) = tanh(x_i^T x_j + b)$	b
Sigmoid	$K(x_i, x_j) = tanh(a x_i^T x_j + b)$	a, b
Bessel function	$K(x_i, x_j) = \dfrac{bessel_{v+1}^n(\sigma\|x_i - x_j\|)}{\left(\|x_i - x_j\|\right)^{-n(v+1)}}$	σ, n, v
ANOVA radial basis	$K(x_i, x_j) = \left(\sum_{k=1}^n e^{-\sigma\left(x_i^k - x_j^k\right)}\right)^d$	σ, d

Table 50: Some kernel functions used in machine learning

kernel function and shows them in Figure 165. The curve illustrates how the similarity *smoothly* decreases when the distance between the two data points increases. And the *bandwidth* parameter γ controls the rate of decrease, i.e., the smaller the γ, the less sensitive the kernel function to the *Euclidean* distance of the data points (measured by $\|x_i - x_j\|^2$).

```r
# Use R to visualize a kernel function
require(latex2exp) # enable the use of latex in R graphics
# write a function for the kernel function
gauss <- function(x,gamma) exp(- gamma * x^2)
x <- seq(from = -3, to = 3, by = 0.001)
plot(x, gauss(x,0.2), lwd = 1, xlab = TeX('$x_i - x_j$'),
     ylab="Gaussian radial basis kernel", col = "black")
lines(x, gauss(x,0.5), lwd = 1, col = "forestgreen")
lines(x, gauss(x,1), lwd = 1, col = "darkorange")
legend(x = "topleft",
       legend = c(TeX('$\\gamma = 0.2$'), TeX('$\\gamma = 0.5$'),
                 TeX('$\\gamma = 1$')),
       lwd = rep(4, 4), col = c("black",
                                "darkorange","forestgreen"))
```

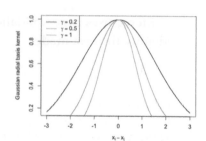

Figure 165: Three instances of the *Gaussian radial basis* kernel function ($\gamma = 0.2$, $\gamma = 0.5$, and $\gamma = 1$)

R Lab

The 6-Step R Pipeline **Step 1** and **Step 2** get the dataset into R and organize it in required format.

```r
# Step 1 -> Read data into R workstation

library(RCurl)
url <- paste0("https://raw.githubusercontent.com",
              "/analyticsbook/book/main/data/KR.csv")
data <- read.csv(text=getURL(url))
# str(data)
```

```
# Step 2 -> Data preprocessing
# Create X matrix (predictors) and Y vector (outcome variable)
X <- data$x
Y <- data$y

# Create a training data
train.ix <- sample(nrow(data),floor( nrow(data) * 4/5) )
data.train <- data[train.ix,]
# Create a testing data
data.test <- data[-train.ix,]
```

Step 3 creates a list of models. For a kernel regression model, important decisions are made on the kernel function and its parameter(s). For example, here, we create two models with two kernel functions and their parameters:

```
# Step 3 -> gather a list of candidate models

# model1: ksmooth(x,y, kernel = "normal", bandwidth=10)
# model2: ksmooth(x,y, kernel = "box", bandwidth=5)
# model3: ...
```

Step 4 uses cross-validation to evaluate the candidate models to identify the best model.

```
# Step 4 -> Use 5-fold cross-validation to evaluate the models

n_folds = 10 # number of fold
N <- dim(data.train)[1]
folds_i <- sample(rep(1:n_folds, length.out = N))

# evaluate model1
cv_mse <- NULL
for (k in 1:n_folds) {
  test_i <- which(folds_i == k)
  data.train.cv <- data.train[-test_i, ]
  data.test.cv <- data.train[test_i, ]
  require( 'kernlab' )
  model1 <- ksmooth(data.train.cv$x, data.train.cv$y,
                  kernel = "normal", bandwidth = 10,
                  x.points=data.test.cv[,1])
  # (1) Fit the kernel regression model with Gaussian kernel
  # (argument: kernel = "normal") and bandwidth = 0.5; (2) There is
  # no predict() for ksmooth. Use the argument
  # "x.points=data.test.cv" instead.
  y_hat <- model1$y
  true_y <- data.test.cv$y
  cv_mse[k] <- mean((true_y - y_hat)^2)
}
```

```
mean(cv_mse)

# evaluate model2 using the same script above
# ...
```

The result is shown below

```
# [1] 0.2605955   # Model1
# [1] 0.2662046   # Model2
```

Step 5 builds the final model.

```
# Step 5 -> After model selection, use ksmooth() function to
# build your final model
kr.final <- ksmooth(data.train$x, data.train$y, kernel = "normal",
                    bandwidth = 10, x.points=data.test[,1]) #
```

Step 6 uses the final model for prediction.

```
# Step 6 -> Evaluate the prediction performance of your model
y_hat <- kr.final$y
true_y <- data.test$y
mse <- mean((true_y - y_hat)^2)
print(mse)
```

This pipeline could be easily extended to KNN smoother model, i.e., using the `knn.reg` in the `FNN` package.

Simulation Experiment   We have created a R script in **Chapter 5** to simulate data from nonlinear regression models. Here, we use the same R script as shown below.

```
# Simulate one batch of data
n_train <- 100
# coefficients of the true model
coef <- c(-0.68,0.82,-0.417,0.32,-0.68)
v_noise <- 0.2
n_df <- 20
df <- 1:n_df
tempData <- gen_data(n_train, coef, v_noise)
```

The simulated data are shown in Figure 166 (i.e., the gray data points).

The following R code overlays the *true* model, i.e., as the black curve, in Figure 166.

```
# Plot the true model
plot(y ~ x, col = "gray", lwd = 2)
lines(x, X %*% coef, lwd = 3, col = "black")
```

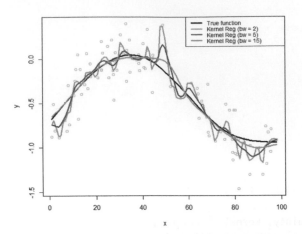

Figure 166: Kernel regression models with different choices on the *bandwidth* parameter (γ) of the Gaussian radial basis kernel function

Let's use the kernel regression model to fit the data. We use the *Gaussian radial basis* kernel function, with three different choices of the *bandwidth* parameter (γ), i.e., ($\gamma = 2$, $\gamma = 5$, $\gamma = 15$). Then we overlay the three fitted kernel regression models in Figure 166 using the following R code.

```
lines(ksmooth(x,y, "normal", bandwidth=2),lwd = 3,
      col = "darkorange")
lines(ksmooth(x,y, "normal", bandwidth=5),lwd = 3,
      col = "dodgerblue4")
lines(ksmooth(x,y, "normal", bandwidth=15),lwd = 3,
      col = "forestgreen")
legend(x = "topright",
       legend = c("True function", "Kernel Reg (bw = 2)",
                  "Kernel Reg (bw = 5)", "Kernel Reg (bw = 15)"),
       lwd = rep(3, 4),
       col = c("black","darkorange","dodgerblue4","forestgreen"),
       text.width = 32, cex = 0.85)
```

As shown in Figure 166, the *bandwidth* parameter determines how smooth are the fitted curves: the larger the bandwidth, the smoother the regression curve[236].

Similarly, we can use the same simulation experiment to study the KNN smoother model. We build three KNN smoother models with $k = 3$, $k = 10$, and $k = 50$, respectively.

[236] Revisit Figure 165 and connect the observations made in both figures, i.e., which one in Figure 165 leads to the smoothest curve in Figure 166 and why?

```
# install.packages("FNN")
require(FNN)
## Loading required package: FNN
xy.knn3<- knn.reg(train = x, y = y, k=3)
xy.knn10<- knn.reg(train = x, y = y, k=10)
xy.knn50<- knn.reg(train = x, y = y, k=50)
```

Similar to Figure 166, we use the following R code to draw Figure 167 that contains the true model, the sampled data points, and the three fitted models.

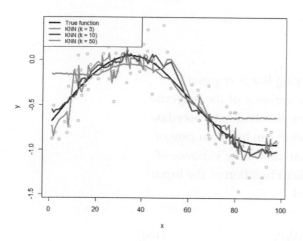

Figure 167: KNN regression models with different choices on the number of nearest neighbors

```
# Plot the data
plot(y ~ x, col = "gray", lwd = 2)
lines(x, X %*% coef, lwd = 3, col = "black")
lines(x, xy.knn3$pred, lwd = 3, col = "darkorange")
lines(x, xy.knn10$pred, lwd = 3, col = "dodgerblue4")
lines(x, xy.knn50$pred, lwd = 3, col = "forestgreen")
legend(x = "topleft",
       legend = c("True function", "KNN (k = 3)",
                  "KNN (k = 10)", "KNN (k = 50)"),
       lwd = rep(3, 4),
       col = c("black", "darkorange", "dodgerblue4",
               "forestgreen"),
       text.width = 32, cex = 0.85)
```

Comparing Figures 166 and 167, it seems that the curve of the kernel regression model is generally *smoother* than the curve of a KNN model. This observation relates to the *discrete* manner the KNN model employs, while the kernel model uses smooth kernel functions that introduce smoothness and continuity into the definition of the neighbors of a data point (thus no hard thresholding is needed to classify whether or not a data point is a neighbor of another data point).

With a smaller k, the fitted curve by the KNN smoother model is less smooth. This is because a KNN smoother model with a smaller k predicts on a data point by relying on fewer data points in the training dataset, ignoring information provided by the other data points that are considered far away[237].

[237] What about a linear regression model? When it predicts on a given data point, does it use all the data points in the training data, or just a few local data points?

In terms of model complexity, the smaller the parameter k in the KNN model, the larger the complexity of the model. Most beginners think of the opposite when they first encounter this question.

Conditional variance regression model

Rationale and formulation

Another common complication when applying linear regression model in real-world applications is that the variance of the response variable may also change. This phenomenon is called **heteroscedasticity** in regression analysis. This complication can be taken care of by a *conditional variance regression* model that allows the variance of the response variable to be a (usually implicit) function of the input variables. This leads to the following model

$$y = \beta^T x + \epsilon_x, \epsilon_x \sim N(0, \sigma_x^2), \tag{104}$$

with ϵ_x modeled as a normal distribution with *varying* variance as a function of x.

The conditional variance regression model differs from the regular linear regression model in terms of how it models σ_x, as illustrated in Figure 168. In other words, the linear regression model is a special type of the conditional variance regression model, with σ_x being a fixed constant as the horizontal line shown in Figure 169. The conditional variance regression model is a stacked model with one regression model on y and another model on σ_x. The model stacking is a common strategy in statistics to handle multi-layered problems.

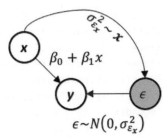

Figure 168: The *data-generating mechanism* of a conditional variance regression model

Theory and method

Given a dataset with N data points and p variables

$$y = \begin{bmatrix} y_1 \\ y_2 \\ \vdots \\ y_N \end{bmatrix}, X = \begin{bmatrix} 1 & x_{11} & x_{21} & \cdots & x_{p1} \\ 1 & x_{12} & x_{22} & \cdots & x_{p2} \\ \vdots & \vdots & \vdots & \vdots & \vdots \\ 1 & x_{1N} & x_{2N} & \cdots & x_{pN} \end{bmatrix}.$$

where $y \in R^{N \times 1}$ denotes the N measurements of the outcome variable, and $X \in R^{N \times (p+1)}$ denotes the data matrix that includes the N measurements of the p input variables and one dummy variable that corresponds to the intercept term β_0. The remaining issue is how to estimate the regression parameters β. There are two situations: σ_x^2 is known and σ_x^2 is unknown.

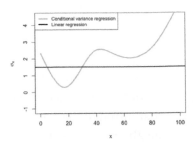

Figure 169: σ_x is a function of x

[238] Readers can use Eq. 104 to derive this likelihood function.

Known σ_x^2   The likelihood function is[238]

$$-\frac{\pi}{2}\ln 2\pi - \frac{1}{2}\sum_{n=1}^{N}\log\sigma_{x_n}^2 - \frac{1}{2}\sum_{n=1}^{N}\frac{\left(y_n - \beta^T x_n\right)^2}{\sigma_{x_n}^2}.$$

As we have known σ_x^2, the parameters to be estimated only involve the last part of the likelihood function. Thus, we estimate the parameters that minimize:

$$\frac{1}{2}\sum_{n=1}^{N}\frac{\left(y_n - \beta^T x_n\right)^2}{\sigma_{x_n}^2}.$$

This could be written in the matrix form as[239]

$$\min_{\beta}(y - X\beta)^T W(y - X\beta). \tag{105}$$

This has the same structure as the *generalized least squares (GLS)* problem we have mentioned in **Chapter 3**. To solve this optimization problem, we take the gradient of the objective function in Eq. 105 and set it to be zero

$$\frac{\partial(y - X\beta)^T W(y - X\beta)}{\partial\beta} = 0,$$

which gives rise to the equation

$$X^T W(y - X\beta) = 0.$$

This leads to the GLS estimator of β

$$\hat{\beta} = (X^T W X)^{-1} X^T W y. \tag{106}$$

Unknown σ_x^2   A more complicated situation, also a more realistic one, is that we don't know σ_x^2. If we can estimate σ_x^2, we can reuse the procedure we have developed for the case when we have known σ_x^2.

As shown in Figure 169, σ_x^2 is a function of x. In other words, it uses the input variables x to predict a new outcome variable, σ_x^2. Isn't this a regression problem? The problem here is we don't have the "measurements" of the outcome variable, i.e., the outcome variable σ_x^2 is not directly measurable[240].

To overcome this problem, we estimate the *measurements* of the latent variable[241], denoted as $\hat{\sigma}_{x_n}^2$ for $n = 1, 2, \ldots, N$. We propose the following steps:

1. Initialize $\hat{\sigma}_{x_n}^2$ for $n = 1, 2, \ldots, N$ by any reasonable approach (i.e., a trivial but popular approach, randomization).

[239] W is a diagonal matrix with its diagonal elements as $W_{nn} = \frac{1}{\sigma_{x_n}^2}$.

[240] Another example of a **latent variable**.

[241] I.e., just like what we did in the EM algorithm to estimate z_{nm}. See **Chapter 6**.

2. Estimate $\hat{\beta}$ using the GLS estimator shown in Eq. 106, and get $\hat{y}_n = \hat{\beta}^T x_n$ for $n = 1, 2, \ldots, N$.

3. Derive the residuals $\hat{\epsilon}_n = y_n - \hat{y}_n$ for $n = 1, 2, \ldots, N$.

4. Build a regression model, e.g., using the kernel regression model, to fit $\hat{e}$ using x.[242]

5. Predict $\hat{\sigma}^2_{x_n}$ for $n = 1, 2, \ldots, N$ using the fitted model in Step 4.

6. Repeat Steps 2 – 5 until convergence or satisfaction of a stopping criteria[243].

> [242] I.e., the training dataset is $\{x_n, \hat{\epsilon}_n, n = 1, 2, \ldots, N\}$.

> [243] E.g., fix the number of iterations, or set a threshold for changes in the parameter estimation.

This approach of taking some variables as latent variables and further using statistical estimation/inference to fill in the unseen measurements has been useful in statistics and used in many models, such as the latent factor models, structural equation models, missing values imputation, EM algorithm, Gaussian mixture model, graphical models with latent variables, etc.

R Lab

Simulation Experiment   We simulate a dataset to see how well the proposed iterative procedure works for the parameter estimation when σ^2_x is unknown. The simulated data has one predictor and one outcome variable. The true model is

$$y = 1 + 0.5x + \epsilon_x, \quad \sigma^2_x = 0.5 + 0.8x^2.$$

We stimulate 100 data points from this model using the R script shown below.

```
# Conditional variance function
# Simulate a regression model with heterogeneous variance
gen_data <- function(n, coef) {
x <- rnorm(100,0,2)
eps <- rnorm(100,0,sapply(x,function(x){0.5+0.8*x^2}))
X <- cbind(1,x)
y <- as.numeric(X %*% coef + eps)
return(data.frame(x = x, y = y))
}
n_train <- 100
coef <- c(1,0.5)
tempData <- gen_data(n_train, coef)
```

The simulated data points are shown in Figure 170, together with the true model (the black line).

To initialize the iterative procedure of parameter estimation for the conditional variance regression model, we fit a regular linear regression model using the following R code.

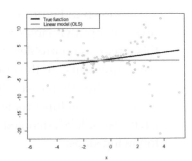

Figure 170: Linear regression model to fit a heteroscedastic dataset

```
# Fit the data using linear regression model (OLS)
x <- tempData[, "x"]
y <- tempData[, "y"]
fit.ols <- lm(y~x,data=tempData)
# Plot the data and the models
x <- tempData$x
X <- cbind(1, x)
y <- tempData$y
plot(y ~ x, col = "gray", lwd = 2)
# Plot the true model
lines(x, X %*% coef, lwd = 3, col = "black")
# Plot the linear regression model (OLS)
lines(x, fitted(fit.ols), lwd = 3, col = "darkorange")
legend(x = "topleft", legend = c("True function",
                              "Linear model (OLS)"),
lwd = rep(4, 4), col = c("black", "darkorange"),
      text.width = 4, cex = 1)
```

The fitted line is shown in Figure 170, which has a significant deviation from the true regression model. We use the fitted line as a starting point, i.e., so that we can estimate the residuals[244] based on the fitted linear regression model. The estimated residuals are plotted in Figure 171 as grey dots. A nonlinear regression model, the kernel regression model implemented by `npreg()` , is fitted on these residuals and shown in Figure 171 as the orange curve. The true function of the variance, i.e., $\sigma_x^2 = 0.5 + 0.8x^2$, is also shown in Figure 171 as the black curve.

It can be seen that the residuals provide a good starting point for us to approximate the underlying true variance function. To reproduce Figure 171, use the following R script.

[244] Residuals $\hat{e}_n = y_n - \hat{y}_n$ for $n = 1, 2, \ldots, N$.

```
# Plot the residual estimated from the linear regression model (OLS)
plot(x,residuals(fit.ols)^2,ylab="squared residuals",
     col = "gray", lwd = 2)
# Plot the true model underlying the variance of the
# error term
curve((1+0.8*x^2)^2,col = "black", lwd = 3, add=TRUE)
# Fit a nonlinear regression model for residuals
# install.packages("np")
require(np)
var1 <- npreg(residuals(fit.ols)^2 ~ x)
grid.x <- seq(from=min(x),to=max(x),length.out=300)
lines(grid.x,predict(var1,exdat=grid.x), lwd = 3,
      col = "darkorange")
legend(x = "topleft",
       legend = c("True function",
                  "Fitted nonlinear model (1st iter)"),
       lwd = rep(4, 4), col = c("black", "darkorange"),
       text.width = 5, cex = 1.2)
```

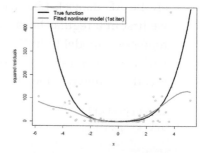

Figure 171: Nonlinear regression model to fit the residuals (i.e., the grey dots)

The orange curve shown in Figure 171 provides an approach to initialize the iterative procedure of parameter estimation for the conditional variance regression model: to estimate the $\hat{\sigma}^2_{x_n}$ for $n = 1, 2, \ldots, N$ in Step 1.[245] Then, for Step 2, we fit a linear regression model according to Eq. 106. This is done by the following R code.

[245] In R, this is done by `fitted(var1)`

```r
# Fit a linear regression model (WLS) with the weights specified
# by the fitted nonlinear model of the residuals
fit.wls <- lm(y~x,weights=1/fitted(var1))
plot(y ~ x, col = "gray", lwd = 2,ylim = c(-20,20))
# Plot the true model
lines(x, X %*% coef, lwd = 3, col = "black")
# Plot the linear regression model (OLS)
lines(x, fitted(fit.ols), lwd = 3, col = "darkorange")
# Plot the linear regression model (WLS) with estimated
# variance function
lines(x, fitted(fit.wls), lwd = 3, col = "forestgreen")
legend(x = "topleft",
       legend = c("True function", "Linear (OLS)",
                  "Linear (WLS) + estimated variance"),
       lwd = rep(4, 4),
       col = c("black", "darkorange","forestgreen"),
       text.width = 5, cex = 1)
```

The new regression model is added to Figure 170 as the green line, which generates Figure 172. The new regression model is closer to the true model.

And we could continue this iterative procedure until a convergence criterion is met.

Real data   Now let's apply the conditional variance regression model on the AD dataset. Like what we did in the simulation experiment, we first fit a regular linear regression model, then, use the kernel regression model to fit the residuals, then obtain the estimates of the variances, then estimate the regression parameters using Eq. 106. The R code is shown below. Results are shown in Figure 173.

```r
library(RCurl)
url <- paste0("https://raw.githubusercontent.com",
              "/analyticsbook/book/main/data/AD.csv")
AD <- read.csv(text=getURL(url))

str(AD)
# Fit the data using linear regression model (OLS)
x <- AD$HippoNV
y <- AD$MMSCORE
fit.ols <- lm(y~x,data=AD)
# Fit a linear regression model (WLS) with the weights specified
# by the fitted nonlinear model of the residuals
```

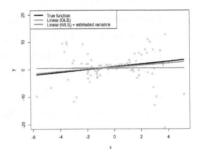

Figure 172: Fit the heteroscedastic dataset with two linear regression models using OLS and GLS (that accounts for the heteroscedastic effects with a nonlinear regression model to model the variance regression)

```
var1 <- npreg(residuals(fit.ols)^2 ~ HippoNV, data = AD)
fit.wls <- lm(y~x,weights=1/fitted(var1))
plot(y ~ x, col = "gray", lwd = 2)
# Plot the linear regression model (OLS)
lines(x, fitted(fit.ols), lwd = 3, col = "darkorange")
# Plot the linear regression model (WLS) with estimated variance
# function
lines(x, fitted(fit.wls), lwd = 3, col = "forestgreen")
legend(x = "topleft",
        legend = c("Linear (OLS)",
                    "Linear (WLS) + estimated variance"),
        lwd = rep(4, 4), col = c("darkorange","forestgreen"),
        text.width = 0.2, cex = 1)
```

We also visualize the fitted variance functions in Figure 174 via the following R code.

```
# Plot the residual estimated from the linear regression
# model (OLS)
plot(x,residuals(fit.ols)^2,ylab="squared residuals",
     col = "gray", lwd = 2)
# Fit a nonlinear regression model for residuals
# install.packages("np")
require(np)
var2 <- npreg(residuals(fit.wls)^2 ~ x)
grid.x <- seq(from=min(x),to=max(x),length.out=300)
lines(grid.x,predict(var1,exdat=grid.x), lwd = 3,
      col = "darkorange")
lines(grid.x,predict(var2,exdat=grid.x), lwd = 3,
      col = "forestgreen")
legend(x = "topleft",
        legend = c("Fitted nonlinear model (1st iter)",
                    "Fitted nonlinear model (2nd iter)"),
                lwd = rep(4, 4),
                col = c( "darkorange", "forestgreen"),
                text.width = 0.25, cex = 1.2)
```

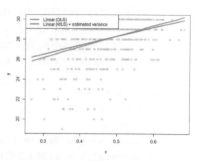

Figure 173: Fit the AD dataset with two linear regression models using OLS and GLS (that accounts for the heteroscedastic effects with a nonlinear regression model to model the variance regression)

Figure 174 shows that in the data the heteroscedasticity is significant. Learning the variance function is helpful in this context. First, in terms of the statistical aspect, it improves the fitting of the regression line. Second, knowing the variance function itself is important knowledge in healthcare, e.g., variance often implies unpredictability or low quality in healthcare operations, pointing out root causes of quality problems or areas of improvement.

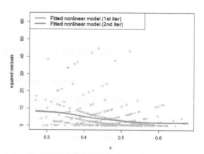

Figure 174: Nonlinear regression model to fit the residuals in the 2nd iteration for the AD data

Remarks

Experiment

The following R code conducts the experiment in Figure 163 (left).

```r
# Write a nice simulator to generate dataset with one
# predictor and one outcome from a polynomial regression
# model
require(splines)
seed <- rnorm(1)
set.seed(seed)
gen_data <- function(n, coef, v_noise) {
  eps <- rnorm(n, 0, v_noise)
  x <- sort(runif(n, 0, 100))
  X <- cbind(1,ns(x, df = (length(coef) - 1)))
  y <- as.numeric(X %*% coef + eps)
  return(data.frame(x = x, y = y))
}
n_train <- 30
coef <- c(1,0.5)
v_noise <- 3
tempData <- gen_data(n_train, coef, v_noise)
tempData[31,] = c(200,200)
# Fit the data using linear regression model
x <- tempData[, "x"]
y <- tempData[, "y"]
fit <- lm(y~x,data=tempData)
# Plot the data
x <- tempData$x
X <- cbind(1, x)
y <- tempData$y
plot(y ~ x, col = "gray", lwd = 2)
lines(x, X %*% coef, lwd = 3, col = "black")
lines(x, fitted(fit), lwd = 3, col = "darkorange")
legend(x = "topleft", legend = c("True function",
       "Fitted linear model"), lwd = rep(4, 4),
       col = c("black", "darkorange"),
       text.width = 100, cex = 1.5)
```

Linear regression as a kernel regression model

Let's consider a simple linear regression problem that has one predictor, x, and no intercept

$$y = \beta x + \epsilon.$$

Given a dataset with N samples, i.e., $\{x_n, y_n, n = 1, 2, \ldots, N.\}$, the least squares estimator of β is

$$\hat{\beta} = \frac{\left(\sum_{i=1}^{N} x_n y_n\right)}{\sum_{n=1}^{N} x_n^2}.$$

Now comes a new data point, x^*. To derive the prediction y^*,

$$y^* = \hat{\beta} x^* = x^* \frac{\left(\sum_{n=1}^{N} x_n y_n\right)}{\sum_{n=1}^{N} x_n^2}.$$

This could be further reformed as

$$y^* = \sum_{n=1}^{N} y_n \frac{x_n x^*}{\sum_{n=1}^{N} x_n^2}.$$

This fits the form of the kernel regression as defined in Eq. 103.[246]

[246] I.e., $w(x_n, x^*) = \sum_{n=1}^{N} \frac{x_n x^*}{\sum_{n=1}^{N} x_n^2}$.

More about heteroscedasticity

For regression problems, the interest is usually in the modeling of the relationship between the *mean*[247] of the outcome variable with the input variables. Thus, when there is heteroscedasticity in the data, a nonparametric regression method is recommended to estimate the latent variance, more from a curve-fitting perspective which is to smooth and estimate, rather than a modeling perspective which is to study the relationship between the outcome variable with input variables. But, of course, we can still study how the input variables affect the variance of the response variable explicitly. Specifically, we can use a linear regression model to link the variance of y with the input variables. The iterative procedure developed for the case when σ_x^2 is unknown is still applicable here for parameter estimation.

[247] See sidenote 11 and Figure 5.

Exercises

ID	x_1	y
1	−0.32	0.66
2	−0.1	0.82
3	0.74	−0.37
4	1.21	−0.8
5	0.44	0.52
6	−0.68	0.97

Table 51: Dataset for building a kernel regression model

1. Manually build a kernel regression model with Gaussian kernel with bandwidth parameter $\gamma = 1$ using the data shown in Table 51, and predict on the data points shown in Table 52.

ID	x_1	y
1	-1	
2	0	
3	1	

Table 52: Testing dataset for the kernel regression model

2. Follow up on the dataset in Q1. Manually build a KNN regression model with $K = 2$. Predict on the testing data in Table 52.

3. Follow up on the dataset in Q1. Use the R pipeline for KNN regression on this data. Compare the result from R and the result by your manual calculation.

4. Follow up on the dataset in Q1. Use the `gausskernel()` function from the R package `KRLS` to calculate the similarity between the data points (including the 6 training data points and the 3 testing data points in Tables 51 and 52).

5. Use the `BostonHousing` dataset from the R package `mlbench`, select the variable `medv` as the outcome, and use other numeric variables as the predictors. Run the R pipeline for KNN regression on it. Use cross-validation to select the best number of nearest neighbor, and summarize your findings.

Figure 175: The true model and its sampled data points

6. Use the `BostonHousing` dataset from the R package `mlbench` and select the variable `lstat` as the predictor and `medv` as the outcome, and run the R pipeline for kernel regression on it. Try the Gaussian kernel function with its bandwidth parameter taking values as $5, 10, 30, 100$.

7. Figure 175 shows a nonlinear model (i.e., the curve) and its sampled points. Suppose that the curve is unknown to us, and our task is to build a KNN regression model with $K = 2$ based on the samples. Draw the fitted curve of this KNN regression model.

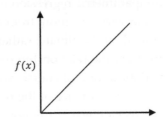

Figure 176: The true model and its sampled data points

8. Suppose that the underlying model is a linear model, as shown in Figure 176. To use KNN model to approximate the underlying model, we need samples. Suppose that we could afford sampling 8 data points. Which locations would you like to acquire samples in order to achieve best approximation of the underlying model using your later fitted KNN model?

Chapter 10: Synthesis
Architecture & Pipeline

Overview

Chapter 10 is about *synthesis*. Synthesis is not a rigorous term but refers to a type of common pratice that integrates, consolidates, or streamlines many otherwise stand-alone models into a mega-model or pipeline. Two styles of synthesis will be introduced here, one represented by deep learning, and another takes the form as pipelines.

Deep learning

To know more about "deep learning", we need to start with its name. The word "deep" is ambiguous but expressive, undetermined but significant. This inviting gesture may have a dazzling effect, but it is based on a specific reason: a deep neural network model is truly deep in terms of its architecture—from input variables to output variables there are many layers in between. Other than that, it is not different from other models in this book. The basic framework of learning as shown in Figure 2 and Eq. 1 in **Chapter 2** still holds true for deep learning.

The word "deep" doesn't imply that other models we have learned so far are not deep. Many models have been studied in great depth, such as the linear models[248] and the learning theory developed for the support vector machine[249]. In this chapter, we will refer to deep learning, specifically to those neural network (NN) models that have many hidden layers, because for NN models we could take the word "deep" at face value—if a model looks deep, it is a deep model. This superficiality, however, builds on a solid foundation[250]: a neural network with a more complex architecture means a more complex form for $f(x)$ in Eq. 1. In other words, this is an attractive proposal, since it suggests we can easily build up depth and capacity of the model by merely increasing its visual complexity. And there have been tools that allow users to drag ready-made modules and piece

[248] Anderson, T. W., *An Introduction to Multivariate Statistical Analysis*, Wiley, 3rd edition, 2003.

[249] Vapnik, V., *The Nature of Statistical Learning Theory*, Springer, 2000.

[250] E.g., the Universal approximation theorem; please refer to Hornik, K., Approximation Capabilities of Multilayer Feedforward Networks, *Neural Networks*, Volume 4, Issue 2, Pages 251-257, 1991.

them together to create the architecture of the NN model they'd like to build, and automatically translate the architecture into its mathematical form and carry out the computational tasks for model training and prediction[251].

[251] E.g., TensorFlow https://www.tensorflow.org/.

Rationale and formulation

An architecture means a function   We have mentioned in **Chapter 2** that the data modeling methods seek explicit forms of $f(x)$ in Eq. 1, while algorithmic modeling methods seek implicit forms. Deep models bend the two. It is like an algorithmic modeling method that you don't need to write up the specific form of $f(x)$, while on the other hand, in theory you could write up $f(x)$ after you have had the architecture[252].

The architecture of a NN model could be quite expressive, i.e., Figure 177 shows an architecture of a neural network model with one layer that is flexible enough to include existing models such as the linear regression model, logistic regression model, and SVM, as shown in Table 53.

[252] In this sense, it is also like the kernel trick used in the SVM model. Remember that in **Chapter 7** we have seen that by using the kernel function in SVM, an implicit transformation of the variables is achieved, and we usually do not know what is the explicit form of $\phi(x)$ the SVM model encodes, but in theory there is such a form of $\phi(x)$.

Figure 177: Architecture of a simple neural network model. The figure is drawn using Alex LeNail's online tool: http://alexlenail.me/NN-SVG/index.html.

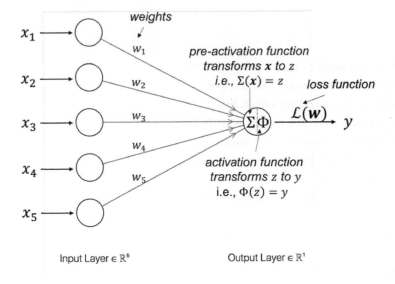

The NN structure shown in Figure 177 is a basic form of NN architecture that is called the **perceptron**. As a basic form, it is a module that could be repeatedly used in different kinds of composition, e.g.,

Model	Activation Function Φ	Loss Function $\mathcal{L}(w)$
Linear regression	Linear: $\Phi(z) = z$	$\mathcal{L}(w) = \left(y - \sum_{i=1}^{p} w_i x_i\right)^2$
Logistic regression	Sigmoid: $\Phi(z) = \frac{1}{1-e^{-z}}$	$\mathcal{L}(w) = \log(1 + \exp[-y \sum_{i=1}^{p} w_i x_i])$
Support vector machine	Null: $\Phi(z) = z$	$\mathcal{L}(w) = \max(0, 1 - y \sum_{i=1}^{p} w_i x_i)$

Table 53: Expression of some models using the architecture of a one-layer neural network in Figure 177

in parallel, concatenation, or in a sequence. The basic forms are also called architectural primitives or foundational building blocks. Most deep architectures are built by combining these architectural primitives. Figures 178 and 179 show two examples. There have been many of those basic forms developed. Softwares such as TensorFlow build on this concept by allowing users to use graphic user interface (GUI) to compose the architecture of their deep networks using these building blocks[253].

As we have mentioned, for NN models there are theories showing that if a model looks deep, it is a deep model. The universal approximation theorem has shown that a NN model with one hidden layer could characterize all smooth functions. While there is no guarantee that in practice adding more layers will always be better, the theoretical results did imply that is the right direction.

[253] For introduction of TensorFlow, readers may check out this book: Ramsundar, B. and Zadeh, R. *TensorFlow for Deep Learning: from Linear Regression to Reinforcement Learning*, O'Reilly Media, 2017.

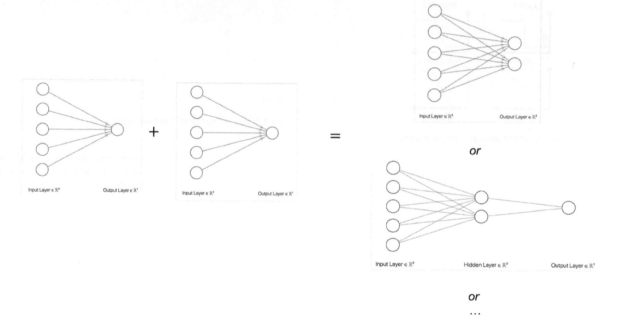

Figure 178: Build more complicated NN models with a basic form

Recall the XOR problem introduced in **Chapter 7** as shown in Figure 123. With a slight modification of the problem to facilitate the presentation here, the dataset has 4 data points

$$x_1 = (0,0), y_1 = 0;$$
$$x_2 = (0,1), y_2 = 1;$$
$$x_3 = (1,0), y_3 = 1;$$
$$x_4 = (1,1), y_4 = 0.$$

This is a typical nonlinear problem. A NN model with one hidden layer as shown in Figure 180 could solve this problem.

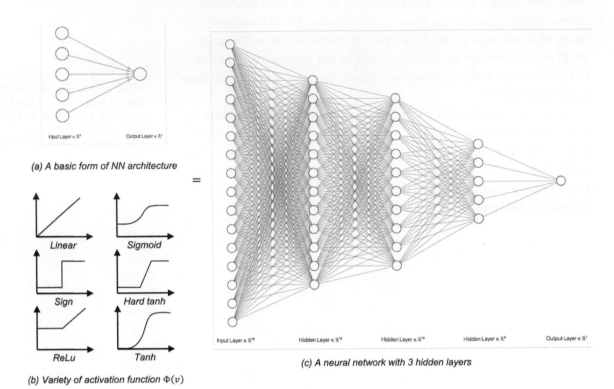

(a) A basic form of NN architecture

Linear

Sigmoid

Sign

Hard tanh

ReLu

Tanh

(b) Variety of activation function $\Phi(v)$

(c) A neural network with 3 hidden layers

Figure 179: Build deeper NN models with basic forms and activation functions

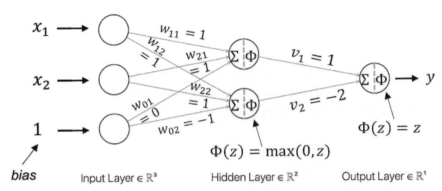

Figure 180: Architecture of a neural network with a hidden layer

For instance, for $x_1 = (0,0)$, from the input layer to the first node (i.e., the upper one) in the hidden layer, we have

$$0 \times 1 + 0 \times 1 + 1 \times 0 = 0.$$

The value 0 provides the input for the activation function at the hidden node, and we have $\Phi(0) = \max(0,0) = 0$.

From the input layer to the second node (i.e., the lower one) in the hidden layer, we have

$$0 \times 1 + 0 \times 1 + 1 \times -1 = -1.$$

The value -1 provides the input for the activation function at the hidden node, and we have $\Phi(-1) = \max(0,-1) = 0$.

Then, from the hidden layer to the output layer, we have

$$1 \times 0 - 2 \times 0 = 0.$$

Using the activation function at the output layer, $\Phi(z) = z$, the final prediction correctly predicts

$$y = 0.$$

We can follow the same process and see that the two-layer NN as shown in Figure 180 could solve the XOR problem.

How to read a deep net   Roughly speaking, there are three major efforts in developing deep learning models: to create basic forms, to design architectural principles or composition rules, and to design learning algorithms that can robustly and efficiently learn the parameters of the deep model using data[254]. Practical application of deep models is to make the network deeper by stacking these basic forms following some composition rules. From this perspective, it is not a surprise to see why it was quoted, "For reason in this sense is nothing but reckoning, that is adding and subtracting … "[255], to explain the logic of designing neural networks in Raul Rojas's book[256].

We can take a look at the convolutional neural networks (CNN) as an example. The CNN is one popular deep NN model and is often used for learning from image data. Its architecture consists of a few basic forms and composition rules that are particularly developed for images.

The CNN architecture shown in Figure 181 has two parts. The first part (i.e., everything before the last 3 layers) is to translate the image data into vectorized form and provides the input for the second part (i.e., the last 3 layers) that is a NN as we have discussed earlier. One basic form of CNN is the convolutional layer. The basic purpose of a

[254] A deep NN model has massive parameters, so learning these parameters from data had been a challenge in the past. Some contributed the recent revitalization of deep learning —as the neural network model had its "rise and fall" in the past decades—to a range of optimization tricks such as pretraining and dropout, the growth of computing power, and the availability of Big Data, all enabled the data-driven learning of a giant collection of parameters of a deep NN model.

[255] Hobbes, T., Leviathan. 1651.

[256] Rojas. R., *Neural Networks: a Systematic Introduction*. Springer, 1996.

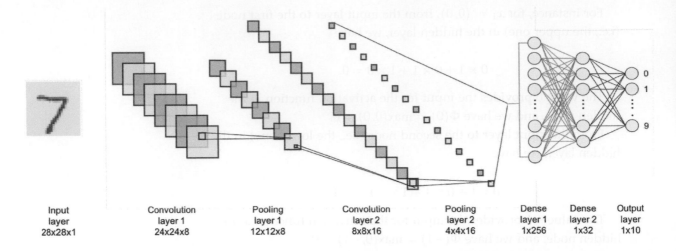

| Input layer 28x28x1 | Convolution layer 1 24x24x8 | Pooling layer 1 12x12x8 | Convolution layer 2 8x8x16 | Pooling layer 2 4x4x16 | Dense layer 1 1x256 | Dense layer 2 1x32 | Output layer 1x10 |

Figure 181: Architecture of a CNN model

convolutional layer is to transform the image into a feature map, as shown in Figure 182.

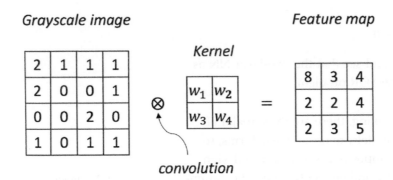

Figure 182: A convolutional layer aggregates spatially correlated information as a feature extraction process

Suppose that $w_1 = 1$, $w_2 = 2$, $w_3 = 2$, $w_4 = 1$ in Figure 182; Figure 183 further shows the computational details of how the convolutional layer works.

The convolutional layer is good at exploiting the spatial structure[257] in its input data. Because of this, CNN is particularly useful for learning from image data, since for images the pixels close to one another are usually semantically related.

The max pooling layer is another basic form of CNN. Figure 184 shows how it works. The max pooling looks too simple an idea, but it works remarkably well. The real mystery when we look at a "simple" idea like this is why it was the max pooling that stood out among many other "simple" ideas. But there has been no conclusive theory to explain it[258].

But one can compare the max pooling with the convolutional layer. One difference is that the parameters of a convolutional layer

[257] I.e., if the entities that are close to each other are semantically related, it is a spatial structure.

[258] To quote Andrew Ng in his online course for convolutional neural networks (https://www.coursera.org/learn/convolutional-neural-networks): "... the main reason people use max pooling is because it's been found in a lot of experiments to work well ... I don't know of anyone who fully knows if that is the real underlying reason."

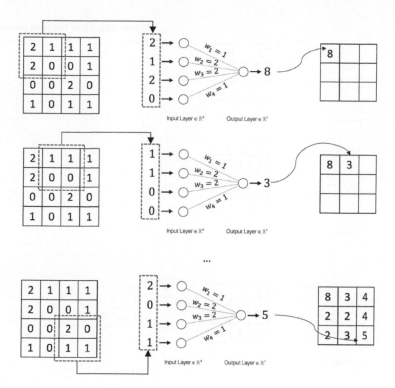

Figure 183: How the convolutional layer works.

is learned from data, making it an adaptive and flexible form to a particular problem. The max pooling, however, is a fixed nonlinear transformation without parameters to learn. In other words, it has no computational cost. No wonder it is believed that one main function of the max pooling is to reduce the number of parameters of the deep NN model and to alleviate the computational cost. This would relieve some computational burden since a deep NN model has a massive number of parameters to be learned from data. Another aspect we should think of is that max pooling is good for image data. It may help increase the robustness of the model against translation invariance, i.e., to recognize an object, say, a cat, in an image, we need the algorithm to be resilient to the potential variation on angle or distance or any other factors that cause scale issues. Max pooling only keeps the "max" and discards the rest.

One can add as many convolutional layers or max pooling layers as needed when designing a CNN model, and the convolutional layers and the max pooling layer could be alternatively arranged as a pipeline to extract features from the image data, e.g., in Figure 181, there are 2 convolutional layers and 1 max pooling layer. It has been found in many cases that for the CNN to be successful, it needs to be made quite deep. For this reason, some consider the deep NN models a different species from NN models.

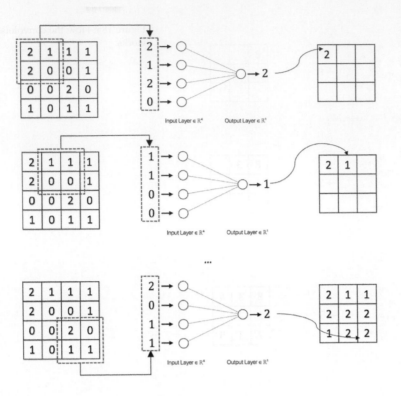

Figure 184: How the max pooling layer works

R Lab

The 6-Step R Pipeline for NN   **Step 1** and **Step 2** get the dataset into R and organize it in required format.

```
# Step 1 -> Read data into R workstation

library(RCurl)
url <- paste0("https://raw.githubusercontent.com",
              "/analyticsbook/book/main/data/KR.csv")
data <- read.csv(text=getURL(url))
# str(data)

# Step 2 -> Data preprocessing
# Create X matrix (predictors) and Y vector (outcome variable)
X <- data$x
Y <- data$y

# Create a training data
train.ix <- sample(nrow(data),floor( nrow(data) * 4/5) )
data.train <- data[train.ix,]
# Create a testing data
data.test <- data[-train.ix,]
```

Step 3 creates a list of models. For a NN model, important decisions are made on the design of the architecture, e.g., how many hidden layers and how many nodes in each hidden layer. For example, here, we create three NN models, all have one hidden layer but a different number of hidden nodes.

```
# Step 3 -> gather a list of candidate models

# NN model with one hidden layer and different # of nodes

# model1: neuralnet(y~x, data=data, hidden=c(3))
# model2: neuralnet(y~x, data=data, hidden=c(5))
# model3: neuralnet(y~x, data=data, hidden=c(8))
```

Step 4 uses cross-validation to evaluate the candidate models to identify the best model.

```
# Step 4 -> cross-validation for model evaluation

n_folds = 10 # number of folds
# the sample size, N, of the dataset
N <- dim(data.train)[1]


folds_i <- sample(rep(1:n_folds, length.out = N))
library(neuralnet)

# cv_mse records the prediction error for each fold
cv_mse <- NULL
for (k in 1:n_folds) {
  # In each iteration of the n_folds iterations
  test_i <- which(folds_i == k)
  # This is the testing data, from the ith fold
  data.test.cv <- data.train[test_i, ]
  # Then, the remaining data form the training data
  data.train.cv <- data.train[-test_i, ]
  # Fit the neural network model with one hidden layer of 3
  model1 <- neuralnet(y~x, data=data, hidden=c(3))
  # Predict on the testing data using the trained model
  pred <- compute (model1, data.test.cv)
  y_hat <- pred$net.result
  model1$y_hat <- y_hat
  # get the true y values for the testing data
  true_y <- data.test.cv$y
  # mean((true_y - y_hat)^2): mean squared error (MSE).
  # The smaller this error, the better your model is
  cv_mse[k] <- mean((true_y - y_hat)^2)
}
mean(cv_mse)
```

The result is shown below

```
# [1] 0.09439574 # Model1
# [1] 0.04433521 # Model2
# [1] 0.1142009  # Model3
```

Obviously, `model2` achieves the lowest prediction error.

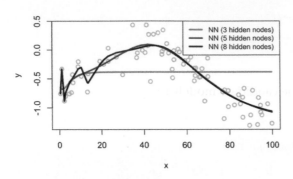

Figure 185: Visualization of the three fitted models and the data

We can also visually examine the fitness of the three models in Figure 185 to see how well the three models fit the data.

```
# Use visual inspection to assist the model selection.

# Predict on the testing data using the trained model
pred <- compute(model1, data.train)
y_hat <- pred$net.result
model1$y_hat <- y_hat
# Predict on the testing data using the trained model
pred <- compute(model2, data.train)
y_hat <- pred$net.result
model2$y_hat <- y_hat
# Predict on the testing data using the trained model
pred <- compute(model3, data.train)
y_hat <- pred$net.result
model3$y_hat <- y_hat


plot(y ~ x, data = data.train, col = "gray", lwd = 2)
lines(data.train$x, model1$y_hat,lwd = 3, col = "darkorange")
lines(data.train$x, model2$y_hat,lwd = 3, col = "blue")
lines(data.train$x, model3$y_hat,lwd = 3, col = "black")
legend(x = "topright", legend = c("NN (3 hidden nodes)",
        "NN (5 hidden nodes)", "NN (8 hidden nodes)"),
     lwd = rep(3, 4), col = c("darkorange", "blue", "black"),
     text.width = 32, cex = 0.85)
```

Step 5 builds the final model. Figure 186 shows the architecture of the final model.

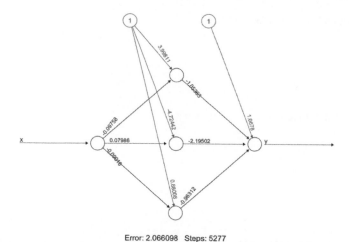

Figure 186: Visualization of the architecture of the final model

```
# Step 5 -> After model selection, build your final model
nn.final <- neuralnet(y~x, data=data.train, hidden=c(5)) #
plot(nn.final) # Draw the architecture of the NN model
```

Step 6 uses the final model for prediction.

```
# Step 6 -> Evaluate the prediction performance of your model
# Predict on the testing data using the trained model
pred <- compute(nn.final, data.test)
y_hat <- pred$net.result
# get the true y values for the testing data
true_y <- data.test$y
# mean((true_y - y_hat)^2): mean squared error (MSE).
# The smaller this error, the better your model is
mse <- mean((true_y - y_hat)^2)
print(mse)
```

The 6-Step R Pipeline for CNN   Before starting the pipeline, let's first install the Keras package.

```
install.packages("devtools") # install devtools
devtools::install_github("rstudio/keras") # install Keras
```

Step 1 and **Step 2** get the MNIST handwritten digit dataset into R and process the data in required format. The goal is to classify a handwritten number into one of the 10 classes (from 0 to 9).

```
# Step 1 -> Read digits classification data
library(keras)
mnist <- dataset_mnist()

# Step 2 -> Data preprocessing
# code adapted from
```

```
# keras.rstudio.com/articles/examples/mnist_cnn.html
# Input image dimensions
img_rows <- 28
img_cols <- 28
num_classes <- 10

# The data, shuffled and split between training and testing sets
x_train <- mnist$train$x
y_train <- mnist$train$y
x_test <- mnist$test$x
y_test <- mnist$test$y

# Redefine  dimension of train/test inputs
x_train <- array_reshape(x_train,
            c(nrow(x_train), img_rows, img_cols, 1))
x_test <- array_reshape(x_test,
            c(nrow(x_test), img_rows, img_cols, 1))
input_shape <- c(img_rows, img_cols, 1)

# Transform RGB values into [0,1] range
x_train <- x_train / 255
x_test <- x_test / 255

cat('x_train_shape:', dim(x_train), '\n')
cat(nrow(x_train), 'train samples\n')
cat(nrow(x_test), 'test samples\n')

# Convert class vectors to binary class matrices
y_train <- to_categorical(y_train, num_classes)
y_test <- to_categorical(y_test, num_classes)
```

Step 3 creates different models. In deep learning, parameters that are determined before training a model are called **hyperparameters**. Hyperparameters for a CNN include number of layers, number of nodes for a layer, kernel size of a convolution layer[259], etc. Here we create three models with different kernel sizes for the convolution layers.

[259] E.g., in Figure 183 the kernel size is 2.

```
# Step 3 -> gather a list of candidate models
define_model <- function(kernel_size){
  model <- keras_model_sequential() %>%
    # convolution layer 1
    layer_conv_2d(filters = 8,
        kernel_size = c(kernel_size,kernel_size),
        activation = 'relu',
        input_shape = input_shape) %>%
    # pooling layer 1
    layer_max_pooling_2d(pool_size = c(2, 2)) %>%
    # convolution layer 2
    layer_conv_2d(filters = 16,
```

```
          kernel_size = c(kernel_size,kernel_size),
          activation = 'relu') %>%
    # pooling layer 2
    layer_max_pooling_2d(pool_size = c(2, 2)) %>%
    # dense layers
    layer_flatten() %>%
    layer_dense(units = 32, activation = 'relu') %>%
    layer_dense(units = num_classes, activation = 'softmax')

  # Compile model
  model %>% compile(
    loss = loss_categorical_crossentropy,
    optimizer = optimizer_adadelta(),
    metrics = c('accuracy')
  )
  return(model)
}
# define three models
model_kernel_1 = define_model(kernel_size=2)
model_kernel_2 = define_model(kernel_size=3)
model_kernel_3 = define_model(kernel_size=5)
```

Step 4 uses cross-validation to evaluate the candidate models to identify the best model.

```
# Step 4 -> Use cross-validation for model evaluation

# set upfunction for evaluating accuracy
cv_accuracy <- function(n_folds, kernel_size,x_train,y_train){
  N <- dim(x_train)[1] # the sample size, N, of the dataset
  folds_i <- sample(rep(1:n_folds, length.out = N))

  accuracy_v <- NULL
  for (k in 1:n_folds) {
    # set up training and testing data
    test_i <- which(folds_i == k)
    x.train.cv <- x_train[-test_i,,,,drop=FALSE]
    x.test.cv <- x_train[test_i,,,,drop=FALSE]
    y.train.cv <- y_train[-test_i,,drop=FALSE ]
    y.test.cv <- y_train[test_i,,drop=FALSE ]

    model <- define_model(kernel_size)
    model %>% fit(
      x_train, y_train, batch_size = 128,
      epochs = 2,validation_split = 0.2, verbose = 0
    )
    scores <- model %>% evaluate(
    x.test.cv, y.test.cv, verbose = 0)

    accuracy_v <- c(accuracy_v, scores[2])
  }
```

```
    return(accuracy_v)
}
# get average accuracy for each model
accuracy_v_kernel_1 <-
    cv_accuracy(n_folds=2,kernel_size=2,x_train,y_train)
print(mean(accuracy_v_kernel_1))

accuracy_v_kernel_2 <-
    cv_accuracy(n_folds=2,kernel_size=3,x_train,y_train)
print(mean(accuracy_v_kernel_2))

accuracy_v_kernel_3 <-
    cv_accuracy(n_folds=2,kernel_size=5,x_train,y_train)
print(mean(accuracy_v_kernel_3))
```

The result is shown below.

```
# [1] 0.9680667 # Model1
# [1] 0.9742167 # Model2
# [1] 0.9760833  # Model3
```

Step 5 builds the final model based on all the training data.

```
# Step 5 -> After model selection, build your final model

model <- define_model(5)
model %>% fit(
    x_train, y_train, batch_size = 128,
    epochs = 2,validation_split = 0.2, verbose = 0
  )
```

Step 6 uses the final model for prediction.

```
# Step 6 -> Evaluate the prediction performance of your model
scores <- model %>% evaluate(
    x_test, y_test, verbose = 0)
print(scores[2])
```

To visualize the process of how this CNN model works, the following R code is used to visualize the output from each layer, shown in Figure 187.

```
# visualize output for a layer

# use the first image from testing data
img <- x_test[1,,,]
plot(as.raster(img))
img <- x_test[1,,,,drop=FALSE]

# define function to plot an image
```

Figure 187: Visualize the outputs from all layers of the CNN model

| Input layer 28x28x1 | Convolution layer 1 24x24x8 | Pooling layer 1 12x12x8 | Convolution layer 2 8x8x16 | Pooling layer 2 4x4x16 | Dense layer 1 1x256 | Dense layer 2 1x32 | Output layer 1x10 |

```r
plot_image <- function(channel) {
    rotate <- function(x) t(apply(x, 2, rev))
    image(rotate(channel), axes = FALSE, asp = 1,
          col = gray.colors(12))
}
# plot the testing image
plot_image(  1 - img[1,,,]   )

# plot the output from the second layer
layer_number = 2

# print layer name
layer_name <- model$layers[[layer_number]]$name
print(layer_name)

layer_outputs <- lapply(model$layers[layer_number],
                        function(layer) layer$output)
activation_model <- keras_model(inputs = model$input,
                                outputs = layer_outputs)
# calculate the outputs from the layer for the image
layer_activation <- activation_model %>% predict(img)

# check dimension
print(dim(layer_activation))

# number of features
n_features <- dim(layer_activation)[[4]]
# image width
image_size <- dim(layer_activation)[[2]]
# number of columns and images per column
# (each column plots an image)
n_cols <- n_features
images_per_col <- 1 #

# plot n_cols of images
op <- par(mfrow = c(n_cols, images_per_col),
          mai = rep_len(0, 4))

# plot each image
for (col in 0:(n_cols-1)) {
        col_ix <- col + 1
        channel_image <- layer_activation[1,,,col_ix]
    plot_image(1-channel_image)
}
```

inTrees

Rationale and formulation

What is a variable? Given an Excel file, we call the column that is ti-
tled as the name *Age* as a variable. And in fact, as a convention, in
an Excel file or a data table in some common formats, we usually do
not doubt that each column implies a variable. These "variables", or
entities, may have definitions in the domain of common sense (i.e.,
where we take things for granted), but they may not be the best candidates to characterize the system under study. Recall that in **Chapter
2** we mentioned that the goal of modeling begins with abstraction—
"identification of a few main entities from the problem", and continues to"characterize their relationships". If we comfortably play the
data using the variables that have been defined without examination,
we lose sight of a large territory of data analytics—identification of a
few main entities from the problem that can sufficiently characterize
the problem.

   Now let's switch to the domain of linear regression. A variable is
an abstract entity in the equation of the linear regression model[260],
multiplied by a regression coefficient[261]. It stands in the equation in
parallel with other variables, which jointly determines an outcome
variable. This form implies that the difference between the variables
are only numeric, characterized by the differences in signs and magnitudes (i.e., encoded in the regression coefficients), but not in semantics. Now comes a reflection: in order for a linear regression model
to work out in an application, shouldn't we ensure that the variables
could be lined up in this manner of apposition[262]?

[260] I.e., denoted as x_i.

[261] I.e., denoted as β_i.

[262] *Apposition*—with a little abuse of the term—the grammar used in the linear regression to line up the variables in a parallel and additive manner.

$$y = \ldots + \underbrace{\beta_1 x_1}_{A} + \underbrace{\beta_2 x_2}_{p} + \underbrace{\beta_3 x_3}_{p} + \underbrace{\beta_4 x_4}_{o} + \underbrace{\beta_5 x_5}_{s} + \underbrace{\beta_6 x_6}_{i} + \underbrace{\beta_7 x_7}_{t} + \underbrace{\beta_8 x_8}_{i} + \underbrace{\beta_9 x_9}_{o} + \underbrace{\beta_{10} x_{10}}_{n} + \ldots$$

$$(107)$$

   In other words, if the variables in the domain of common sense are
not semantic equals, the application of linear regression on them is
questionable. For example, it is probably common sense to line up a
few genetic factors in Eq. 107, but could we also put *Age* as another
variable that stands among the genetic factors in a line? In many
contexts, we need to work out a better definition of the variable, i.e.,
it is not uncommon to define two new variables such as $Age \leq 65$
and $Age > 65$ instead of using the variable *Age* directly. Sometimes
we use the variable *Age* in a model because it is named as *Age*. But
what is *Age*? When we put a variable in a model, it is destined to be
*re*defined, either before the analysis, or after, or along the way.

   This effort to redefine variables could be automated by tree models. Recall that the tree models use rule-based semantics. Rules like *Age* ≤ 65 and *Age* > 65 sometimes yield statistically significant *and* semantically meaningful entities, perfect candidates for variable redefinition purposes. And a tree is essentially a collection of multiple rules. If we could run tree models on a dataset first, we could extract those rules, and each rule is a new variable.

This is the starting point of `inTrees` . It uses the random forest to collect potentially useful rules[263], then puts the rules as the variables into a model[264] and employs a computational process to select the final variables. In this way, we have the best parts of both methods: the rules capture the variable-level patterns in the data, and the model (i.e., a regression model) captures the synthetic effects of these patterns in predicting an outcome variable. Note that `inTrees` is not the first of its kind. It follows a few pioneers such as the `rulefit` [265] and modifies existing efforts according to some in-field experiences.

[263] This step automates the variable redefinition process.

[264] I.e., a classification/regression model. This step automates the integration of the variable redefinition with modeling.

[265] Friedman, J.H. and Popescu, B.E., *Predictive learning via rule ensembles*. Annals of Applied Statistics, Volume 2, Number 3, Pages 916-954, 2008.

Theory and method

The `inTrees` uses a framework that is shown in Figure 188. In the following text, we introduce each functionality of the `inTrees` framework.

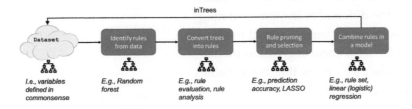

Figure 188: The pipeline of `inTrees`

Consider the dataset that has 2 predictors and 7 instances as shown in Table 54.

ID	x_1	x_2	Class
1	1	1	C0
2	1	0	C1
3	0	1	C1
4	0	0	C1
5	0	0	C0
6	0	0	C0
7	0	0	C0

Table 54: Example of a dataset with 7 instances

Extract rules The `inTrees` uses a tree emsemble learning method to grow many trees. A decision tree can be dissembled into a set of rules. For example, suppose that a random forest model has been built on the dataset shown in Table 54. One tree of this random forest model is shown in Figure 189. Three rules (each rule corresponds to a leaf node) are extracted and shown in Table 55.

Each rule is evaluated by three criteria: the **length** of a rule that is defined as the number of variables in the rule; the **frequency** of a rule that is the proportion of data points in the dataset that meet the condition of the rule, and the **error rate** of a rule. For classification problems, the error rate of a rule is the number of data points incorrectly identified by the rule divided by the number of data points that meet the condition of the rule.

For regression problems, the error rate of a rule is the *mean squared error (MSE)*, that is defined as

$$MSE = \frac{1}{N} \sum_{i=1}^{N} (y_i - \bar{y})^2,$$

where N is the number of data points in the leaf node that corresponds to the rule, y_i is the value of the outcome variable of the ith data point, and $\bar{y}$ is the average of the outcome variable (i.e., as the prediction at the leaf node).

Based on these three criteria, the evaluation of the three rules is shown in Table 55.[266]

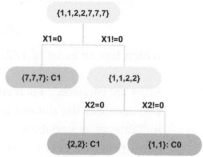

Figure 189: Example of a decision tree; leaf nodes (a.k.a., decision nodes) are shadowed in gray.

[266] Apply each rule on the data points in Table 54.

ID	Rule	Length	Frequency	Error
1	$\{x_1 = 0 \rightarrow Class = C_0\}$	1	5/7	2/5
2	$\{x_1 \neq 0, x_2 = 0 \rightarrow Class = C_1\}$	2	1/7	0/1
3	$\{x_1 \neq 0, x_2 \neq 0 \rightarrow Class = C_0\}$	2	1/7	0/1

Table 55: Evaluation of the three rules extracted from the tree in Figure 189

Prune rules   A lengthy rule, i.e., a rule with many variables, is hard to interpret. For example, consider a rule

Rule: $\{Age \leq 65, Gene\ A = Type\ 1, Gene\ B = Type\ 3 \rightarrow Class = No\ risk\}$.

It is unknown if the three variables are equally important in making the prediction. And, because the way the random forests grow the trees, it is possible that some variables in a rule are not significant at all[267]. Therefore, it is beneficial to prune the rules and remove irrelevant variables from the rules.

Take Rule 2 $\{x_1 \neq 0, x_2 = 0 \rightarrow Class = C_1\}$ for example. The error rate for this rule is 0.[268] Now remove $x_1 \neq 0$ from the rule, and the new rule becomes

$$\{x_2 = 0 \rightarrow Class = C_1\},$$

which has an error of 3/5.[269] Therefore, the error rate increases by 3/5. This increase of error rate is named *decay* in the terminology of

[267] I.e., these variables are selected because the random forest model purposely *randomizes* the learning process.

[268] See Table 55.

[269] Use Table 54.

`inTrees` . A threshold is set by the user, i.e., here, if the threshold is set to be 0.05, we should not remove x_1 from Rule 2 since $3/5 > 0.05$.

Now let's remove $x_2 = 0$. The resulting rule is

$$\{x_1 \neq 0 \rightarrow Class = C_1\},$$

which has an error of $1/2$. Therefore, x_2 should not be pruned either.

Rules are variables   Each rule leads to a redefined variable. For example, consider the dataset in Table 54. We name the three rules shown in Table 55 as variables z_1, z_2, and z_3, respectively. We only use the **condition of a rule** to define the variable. For instance, the condition of a rule is illustrated below

$$\underbrace{\{x_1 \neq 0}_{condition} \rightarrow \underbrace{Class = C_1\}}_{outcome}.$$

Consider z_1 first[270]. The data points in Table 54 that meet the condition $\{x_1 = 0\}$ include $ID = \{3, 4, 5, 6, 7\}$. Thus, the values of z_1 are $\{0, 0, 1, 1, 1, 1, 1\}$. For z_2,[271] only data point $ID = \{2\}$ meets the condition, and therefore, the values of z_2 are $\{0, 1, 0, 0, 0, 0, 0\}$. Similarly, the values of z_3 are $\{1, 0, 0, 0, 0, 0, 0\}$.[272]

The new dataset is shown in Table 56.

[270] I.e., $\{x_1 = 0\}$.

[271] I.e., $\{x_1 \neq 0, x_2 = 0\}$.

[272] I.e., $\{x_1 \neq 0, x_2 \neq 0\}$.

ID	z_1	z_2	z_3	Class
1	0	0	1	C_0
2	0	1	0	C_1
3	1	0	0	C_1
4	1	0	0	C_1
5	1	0	0	C_0
6	1	0	0	C_0
7	1	0	0	C_0

Table 56: The *binarized* dataset of Table 54 by the rules in Table 55

Select rules   A feature selection method could be applied on the new dataset to select the significant variables. Methods such as the L_1 regularized logistics regression (i.e., the equivalent of LASSO for logistic regression model) and regularized random forests are used in the `inTrees` .

Note that most existing methods don't concern the length of the rules. But, given two rules with the same predictive power, the rule with a shorter length should be preferred[273]. In `inTrees` , the **Guided Regularized Random Forest (GRRF)** is also an option for feature selection: the GRRF can assign a weight to each variable, so that when two variables have similar predictive power, the variable with higher weight is more likely to be selected. In our case, we could set higher weight[274] for shorter rules in GRRF.

[273] A shorter rule means a simpler model, better interpretability, etc.

[274] What is the optimal weight? We could use cross-validation to decide.

Rule ensemble As the rules are taken as new variables, a new dataset such as the one shown in Table 56 is created. So theoretically, any model could be applied on the new dataset to build a prediction model. There are preferences in different packages. For example, in `RuleFit`, a linear regression model is used that takes the rules as predictors. In `inTrees`, a simple rule ensemble method summarizes the rules into an *ordered rule set* for prediction.

It takes a few iterations to develop the ordered rule set. First, we create a default rule, denoted as r_0, that has a null *condition* and classifies all the data points to be the most frequent class (if it is a regression model, then r_0 predicts all the data points to be the population average).

Denote the ordered rule set as R, which is set to be empty at the beginning. Then, the algorithm searches through the available rules and identifies the best rule and adds it into R. The best rule is defined as the rule with the minimum error evaluated by the training data. If there are ties, the rule with higher frequency and smaller length is selected. Then, the data points that meet the condition of the best rule are removed, and the default rule r_0 is re-calculated with the data points left. The algorithm iterates to search for the next best rule and update r_0 after each iteration. This iterative process continues until no data point is left in the training dataset, or the default rule r_0 beats all other available rules that have not been added into the rule ensemble set R. Note that, the selected rules in R are ordered according to the sequential order of their inclusion.

Consider the dataset shown in Table 54 and the rules shown in Table 55. The error rate and frequency of each rule is shown in Table 57.

ID	Rule	Error	Frequency
0	$\{Class = C_0\}$	3/7	7/7
1	$\{x_1 = 0 \rightarrow Class = C_0\}$	2/5	5/7
2	$\{x_1 \neq 0, x_2 = 0 \rightarrow Class = C_1\}$	0/1	1/7
3	$\{x_1 \neq 0, x_2 \neq 0 \rightarrow Class = C_0\}$	0/1	1/7

Table 57: Error rates and frequencies of the rules in Table 55 using the dataset in Table 54

At the beginning, the default rule is $\{Class = C_0\}$, and its error rate is 3/7. The algorithm then searches for the best rule in the available rules (i.e., shown in Table 55). Rule 2 and Rule 3 have the least errors, and their frequency and length are also the same. Thus, we can add either of them into R. Assume that Rule 2 is selected: $R = \{x_1 \neq 0, x_2 = 0 \rightarrow Class = C_1\}$. Then, the data point (ID:2) classified by this rule is removed from Table 54. The default rule r_0 is still $\{Class = C_0\}$, and the error and frequency of each rule on the updated dataset[275] is updated, as shown in Table 58.

A new iteration begins and $\{x_1 \neq 0, x_2 \neq 0 \rightarrow Class = C_0\}$ is

[275] The data point (ID:2) is removed.

ID	Rule	Error	Frequency
0	$\{Class = C_0\}$	2/6	6/6
1	$\{x_1 = 0 \to Class = C_0\}$	2/5	5/6
2	$\{x_1 \neq 0, x_2 = 0 \to Class = C_1\}$	NA	0/6
3	$\{x_1 \neq 0, x_2 \neq 0 \to Class = C_0\}$	0/1	1/6

Table 58: Updated error rates and frequencies of the rules in Table 55 using the reduced dataset, i.e., data point (ID:2) in Table 54 is removed

added to R, and the data point (ID:1) is removed. The default rule remains unchanged and the error and frequency of each rule on the updated dataset[276] is updated in Table 59.

[276] The data points (ID:1 and ID:2) in Table 54 are removed.

ID	Rule	Error	Frequency
0	$\{Class = C_0\}$	2/5	5/5
1	$\{x_1 = 0 \to Class = C_0\}$	2/5	5/5
2	$\{x_1 \neq 0, x_2 = 0 \to Class = C_1\}$	NA	0/5
3	$\{x_1 \neq 0, x_2 \neq 0 \to Class = C_0\}$	NA	0/5

Table 59: Updated error rates and frequencies of the rules in Table 55 using the reduced dataset, i.e., data points (ID:1 and ID:2) in Table 54 are removed

Now the default rule $Class = C_0$ has the minimum error 2/5, the same as $\{x_1 = 0 \to Class = C_0\}$. Therefore, the default rule is added to R and the process stops. The final ordered rule set R is summarized in Table 60.

Order	Rule
0	$\{Class = C_0\}$
1	$\{x_1 \neq 0, x_2 = 0 \to Class = C_1\}$
2	$\{x_1 \neq 0, x_2 \neq 0 \to Class = C_1\}$

Table 60: Final results of R

When predicting on an instance, the first rule in R that *hits* the data point is used for prediction. For example, for a data point $\{x_1 \neq 0, x_2 = 1\}$, it meets the condition of Rule 2 in R. The prediction on this data point is C_1. For data point $\{x_1 = 0, x_2 = 1\}$, it does not meet the condition of either Rule 1 or Rule 2 in R. Therefore, the default rule is used, and the prediction is C_0.

R Lab

We use `inTrees` on the AD datasedt. Based on the random forest model, 4555 rules are extracted.

```
rm(list = ls(all = TRUE))
library("arules")
library("randomForest")
library("RRF")
library("inTrees")
library("reshape")
library("ggplot2")
set.seed(1)
url <- paste0("https://raw.githubusercontent.com",
```

```
            "/analyticsbook/book/main/data/AD.csv")
data <- read.csv(text=getURL(url))

target_indx <- which(colnames(data) == "DX_bl")
target <- paste0("class_", as.character(data[, target_indx]))
rm_indx <- which(colnames(data) %in%
            c("DX_bl", "ID", "TOTAL13", "MMSCORE"))
X <- data
X <- X[, -rm_indx]
for (i in 1:ncol(X)) X[, i] <-
        as.factor(dicretizeVector(X[, i], K = 3))

## Use random forests to grow the trees
rf <- randomForest(X, as.factor(target))

# transform rf object to an inTrees' format
treeList <- RF2List(rf)
exec <- extractRules(treeList, X)  # Extract the rules

## The rules are measured by length, error and frequency.
class <- paste0("class_", as.character(target))
rules <- getRuleMetric(exec, X, target)
```

The statistics of the rules could be extracted, e.g., 5 rules are shown below.

```
print(rules[order(as.numeric(rules[, "len"])), ][1:5, ])

#       len freq    err
# [1,] "2" "0.118" "0.098"
# [2,] "2" "0.182" "0"
# [3,] "2" "0.182" "0"
# [4,] "2" "0.081" "0.024"
# [5,] "2" "0.043" "0.136"
#       condition                                       pred
# [1,] "X[,6] %in% c('L1') & X[,11] %in% c('L1')"    "class_1"
# [2,] "X[,4] %in% c('L1') & X[,6] %in% c('L1')"     "class_1"
# [3,] "X[,4] %in% c('L1') & X[,6] %in% c('L1')"     "class_1"
# [4,] "X[,3] %in% c('L3') & X[,4] %in% c('L3')"     "class_0"
# [5,] "X[,6] %in% c('L3') & X[,7] %in% c('L3')"     "class_0"
```

We then prune the rules. Recall that we need to specify the threshold of *decay*. This could be done in R by setting the value of the parameter `maxDecay`. The statistics of the rules before and after pruning are shown in Figures 190—192.

The R code below generates Figure 190.

```
rules.pruned <- pruneRule(rules, X, target, maxDecay = 0.005,
                          typeDecay = 2)
```

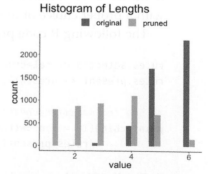

Figure 190: Histogram of *lengths of the rules* before and after the pruning

```
length <- data.frame(original = as.numeric(rules[, "len"]),
                pruned = as.numeric(rules.pruned[,"len"]))

## Visualize the result
ggplot(melt(length), aes(value, fill = variable)) +
  geom_histogram(position = "dodge",binwidth = 0.4) +
  ggtitle("Histogram of Lengths") +
  theme(plot.title = element_text(hjust = 0.5))
```

The R code below generates Figure 191.

```
frequency <- data.frame(
                original = as.numeric(rules[, "freq"]),
                pruned = as.numeric(rules.pruned[,"freq"]))
ggplot(melt(frequency), aes(value, fill = variable)) +
  geom_histogram(position = "dodge",binwidth = 0.05) +
  ggtitle("Histogram of Frequencies") +
  theme(plot.title = element_text(hjust = 0.5))
```

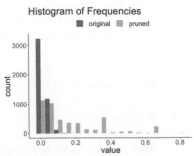

Figure 191: Histogram of *frequencies of the rules* before and after the pruning

The R code below generates Figure 192.

```
error <- data.frame(original = as.numeric(rules[, "err"]),
                pruned = as.numeric(rules.pruned[,"err"]))

## Visualize the result
ggplot(melt(error), aes(value, fill = variable)) +
  geom_histogram(position = "dodge",binwidth = 0.01) +
  ggtitle("Histogram of Errors") +
  theme(plot.title = element_text(hjust = 0.5))
```

Figure 190 shows that the lengths of the rules are substantially reduced. For example, a majority of the original rules have a length of 6, while after pruning, only a slight percentage of the rules have a length of 6. Also, since rules are shortened, reduction of frequencies of the rules is also significant, as shown in Figure 191. The errors are also reduced; e.g., Figure 192 shows the distribution of errors shifted to the left after pruning. Overall, the quality of the rules is improved with a proper choice of the pruning parameters.

The following R code prunes the rule set.

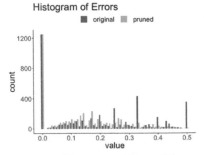

Figure 192: Histogram of *errors of the rules* before and after the pruning

```
rules.selected <- selectRuleRRF(rules.pruned, X, target)
rules.present <- presentRules(rules.selected, colnames(X))

## See the specific contents of the selected rules
print(cbind(ID = 1:nrow(rules.present),
            rules.present[, c("condition", "pred")]))
```

Finally, 16 rules are selected. Their performances are shown below[277].

[277] Details of the rules could also be printed out in R.

```
print(cbind(ID = 1:nrow(rules.present),
            rules.present[, c("len", "freq", "err")]))
##       ID   len  freq     err
## [1,] "1"  "2"  "0.279" "0.083"
## [2,] "2"  "2"  "0.279" "0.09"
## [3,] "3"  "5"  "0.029" "0.133"
## [4,] "4"  "3"  "0.122" "0.016"
## [5,] "5"  "4"  "0.031" "0.312"
## [6,] "6"  "2"  "0.207" "0.121"
## [7,] "7"  "3"  "0.172" "0.124"
## [8,] "8"  "4"  "0.06"  "0.194"
## [9,] "9"  "5"  "0.006" "0"
## [10,] "10" "4"  "0.044" "0.13"
## [11,] "11" "5"  "0.019" "0.2"
## [12,] "12" "3"  "0.043" "0.182"
## [13,] "13" "4"  "0.037" "0.158"
## [14,] "14" "3"  "0.114" "0.203"
## [15,] "15" "2"  "0.234" "0.215"
## [16,] "16" "3"  "0.282" "0.144"
```

Remarks

Images, text, and audio

To learn more about deep learning we recommend readers to start
with this book[278]. There are many online lecture notes and tutorials
that are informative. Here, it is worth mentioning that there has not
been a unified theory about deep learning, and even the definition
of what is a deep model is up to debate. This is good. If we look
back at the developmental processes of many statistics and machine
learning models, we may observe that some models were developed
based on inspiration from theory and we often call these models
too theoretical. These models usually wobble and stumble in their
early years, gradually become mature and a proven approach, and
eventually establish themselves as effective models in practice. Some
other models, however, were developed ahead of theory, and theory
only comes later to explain the model's success. For deep learning, it
is hard to say if it was theory that inspired the models, or it was the
models that inspired theory. Many efforts are committed to give an
overarching theory to explain the success of deep learning, at least in
some special cases. Yet there has not been such an overarching theory
about deep learning, only competing narratives[279].

It is natural to wonder why we use a deep model. Can't we just
use a nondeep model? Readers may have been using nondeep models and found those nondeep models sufficient to solve problems in
practice. Certainly we can just use nondeep models. There have been

[278] Goodfellow, I., Bengio, Y., and
Courville, A., *Deep Learning*. The MIT
Press, 2016.

[279] Interested readers may read this
article by Colah, C., *Neural Networks,
Types, and Functional Programming*,
https://colah.github.io/posts/
2015-09-NN-Types-FP/.

plenty of examples in practice that nondeep models were the best[280], only if we have the best variables (e.g., the $x_1, x_2, \ldots, x_p$) that are sufficient to explain the "movement" of our target y. The availability of high-quality and ready-to-use x is a precondition for the success of nondeep models. This precondition, however, is not always held in practice.

This is one reason why deep models can make a difference. Most nondeep models deal with a data structure that is Excel-sheet-like, i.e., they are stored or could be stored in an Excel spreadsheet. In many applications, particularly in recent years, the raw data is in free-form (sometimes it is also called unstructured data) such as images, text, and audio data. That means, to use the nondeep models for these applications, there should be a preprocessing/translational step that could extract the variables x from the raw data. It is notable that the translational process itself takes up a larger portion of effort of a data scientist in practice, and since the process involves multiple steps and layers, it naturally adopts a deep form, and further includes the nondeep model as its last layer to be part of its architecture.

It is no surpise then that mature practices of deep learning have been mainly found on unstructured data such as images, text, and audio data. From the raw data such as an X-ray image to the final outcome such as diagnosis of a disease, there are plenty of steps to transform the raw data into interpretable information. These steps put together creates a deep model. In other words, deep learning automates and optimizes this translational process.

A key is made to unlock, but what is the lock?

There is another magic dimension to deep learning. Researchers in different disciplines have created many basic forms of functions that could be used as building blocks to build larger functions. On the other hand, as model validation has been made automatic and data-driven[281], a deep model doesn't demand interpretability or validity to be useful. In practice, your task is empirical: put together the architecture of a deep model, and if it obtains superior performance on data, it is a superior model. Now, if a complex problem in the real world is a sophisticated lock, deep learning's real appeal is that we only need to spend effort in guessing at the key (i.e., the architecture of the deep neural network), but not necessarily in understanding the lock. And what makes it more convenient is that we can try every key we made (i.e., by fitting it with data) until the lock is opened. Is this a rational practice? There have been discussions around this topic and readers may be interested to look into this[282].

[280] A recent example: The Math of March Madness, *New York Times*, https://www.nytimes.com/2015/03/22/opinion/sunday/making-march-madness-easy.html.

[281] I.e., given a few candidate models, we no longer need to validate the models by their scientific implication but only check how well they fit the data. See Figure 94 in **Chapter 5**.

[282] See, Hutson, M., Has artifical intelligence become alchemy?, *Science*, 2018.

Decay and relative decay

To prune a rule, `inTrees` uses *leave-one-out* pruning, i.e., at each round, it removes one variable and checks how much error this removal will induce. We have introduced the concept *decay*. For the ith variable in the condition of a rule, its decay is defined as

$$decay_i = Err_{-i} - Err,$$

where Err is the error of the original rule, Err_{-i} is the error of the rule with the ith variable removed.

There is another definition of decay in `inTrees`, called *relative decay*, which is defined as

$$decay_i = \frac{Err_{-i} - Err}{\max(Err, s)},$$

where s is a small positive constant (e.g., 0.001) that bounds the value of decay when Err is zero or close to zero. An advantage of using relative decay is that one does not need to know the level of error of a dataset[283].

[283] For instance, for one dataset the error rate 0.01 is probably insignificant, but for another 0.01 is a big difference.

Exercises

1. Complete the convolution operation as shown in Figure 193.

2. Use the `convolution()` function in R package `OpenImageR` to run the data in Q1.

3. Let's try applying the convolution operation on a real image. For example, use the following R code to get the image shown in Figure 194.

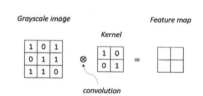

Figure 193: A NN model with its parameters

```
library(EBImage)
readImage(system.file("images", "sample-color.png",
img <- readImage(system.file("images", "sample-color.png",
                 package="EBImage"))
grayimage<-channel(img,"gray")
display(grayimage)
```

Use the `convolution()` function in R package `OpenImageR` to filter this image. You can use the high-pass Laplacian filter, that would be defined in R as

Figure 194: Data for Q3

```
kernel = matrix(1, nc=3, nr=3)
kernel[2,2] = -8
```

4. Figure 195 shows a NN model with its parameters. Use this NN model to predict on the data points shown in Table 61.

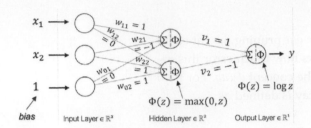

ID	x_1	x_2	y
1	0	1	
2	−1	2	
3	2	2	

Table 61: Test dataset for the NN model in Q4

5. Use the `BostonHousing` dataset from the R package `mlbench` and select the variable `medv` as the outcome and all other numeric variables as predictors. Run the R pipeline for NN on it. Use 10-fold cross-validation to evaluate a NN model with 2 hidden layers, while each layer has a number of nodes of your choice. Comment on the result.

Conclusion

In his book[284] published in 1939—the era when statistics and mass production found each other—Walter A. Shewhart wrote, *"In any case ... an observed sequence is or is not random can be verified only in the future"*. The sequence of observations shown in Figure 37 in **Chapter 3** must have puzzled many in his time, and we know it was a time of chaos. The *data* without being looked at by a model is like a mountain in the Cascade Range that has not been named. Why, according to Ayn Rand, do the machines have their moral code *"Every part of the motors was an embodied answer to 'Why?' and 'What for?'"*[285]. Not surprisingly, but still conceptually challenging even nowadays for newcomers to accept his idea, Walter A. Shewhart created the concept of a control chart to frame the sequence of observations in Figure 37. His brilliant and bold idea is shown in Figure 38.

We are still doing the same thing Shewhart had done to create a *frame* and look at the data *within* the frame, as Figure 196 shows.

Sometimes, we even love the frame more than the picture. But any valuable effort in practice must meet a specific goal, and how the frame can help us reach the goal is the first step to exercise our knowledge, willpower, and judgment. So, without further rambling, we invite you to answer this question: how do you connect the 9 points in Figure 196 with 4 lines?

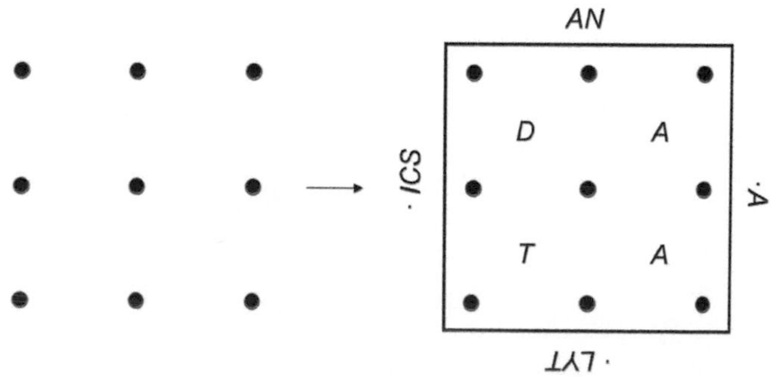

[284] Shewhart, W.A., *Statistical Method from the Viewpoint of Quality Control*, The Department of Agriculture, 1939.

[285] Rand, A., *Atlas Shrugged*, Random House, 1957.

Figure 196: (Left) the *data*; (right) the *data* being looked at *within* a frame

Readers, congratulations. It is time to turn the page for the answer in Figure 197.

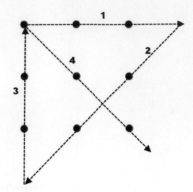

Figure 197: The answer to connecting 9 points with 4 lines

Appendix: A Brief Review of Background Knowledge

Recall that in this book, we use lower case letters, e.g., x, to represent scalars; bold face, lower case letters, e.g., $\mathbf{x}$, to represent vectors; and bold face, upper case letters, e.g., $\mathbf{X}$, to represent matrices.

The Normal Distribution

A distribution model characterizes the random behavior of a random variable. A random variable takes value from a predefined set, range, or a continuum, but not all values are taken with equal probabilities. How these probabilities are distributed is characterized by the distribution model. Before the computer age, for a distribution model to acquire a status of natural law it usually has an elegant geometric shape that comes with a delicate mathematical form, as many examples shown in Figure 4. As we have computers now doing a lot of computation, a distribution could be just an empirical histogram that has not yet found its explicit mathematical form. Whether or not this empirical form of distribution would repeat itself as a natural law remains to be seen. In practice, a competitive edge could be gained before you find scientific explanation, as long as it works.

In this book we will not have extensive coverage of distribution models. We will focus on normal distribution; but other than that, everything we learn about the normal distribution is also to help us extend beyond it and establish the concept of distribution as an abstract one.

A random variable x distributed as a normal distribution is denoted as

$$x \sim N\left(\mu, \sigma^2\right).$$

The normal distribution has the mathematical form

$$N\left(\mu, \sigma^2\right) = \frac{1}{\sqrt{2\pi}\sigma} e^{-\frac{1}{2}\left(\frac{x-\mu}{\sigma}\right)^2}.$$

If we multiply x with a constant a, then

$$ax \sim N\left(a\mu, a^2\sigma^2\right).$$

Extending the concept of distribution to p-dimensional space, we have the multivariate normal distribution (MVN) of vector x

$$x \sim MVN\left(\mu, \Sigma\right),$$

where

$$\mu = \begin{bmatrix} \mu_1 \\ \mu_2 \\ \vdots \\ \mu_p \end{bmatrix}, \quad \Sigma = \begin{bmatrix} \sigma_1^2 & \sigma_{12} & \cdots & \sigma_{1p} \\ \sigma_{21} & \sigma_2^2 & \cdots & \sigma_{2p} \\ \vdots & \vdots & \vdots & \vdots \\ \sigma_{p1} & \sigma_{p2} & \cdots & \sigma_p^2 \end{bmatrix},$$

and

$$MVN\left(\mu, \Sigma\right) = \frac{1}{\sqrt{(2\pi)^p \det \Sigma}} \exp\left(-\frac{1}{2}(x-\mu)^T \Sigma^{-1}(x-\mu)\right).$$

To interpret the covariance matrix Σ, let's look at an example where $p = 2$. Its covariance matrix is

$$\Sigma_1 = \begin{bmatrix} \sigma_1^2 & \sigma_{12} \\ \sigma_{21} & \sigma_2^2 \end{bmatrix}.$$

The element σ_1^2 is the marginal variance of variable x_1, σ_2^2 is the marginal variance of variable x_2, and σ_{12} that equals to σ_{21} is the covariance between x_1 and x_2.[286]

Three examples of the covariance matrix are shown below

$$\Sigma_1 = \begin{bmatrix} 1 & 0 \\ 0 & 1 \end{bmatrix}, \quad \Sigma_2 = \begin{bmatrix} 1 & 0.8 \\ 0.8 & 1 \end{bmatrix}, \quad \Sigma_3 = \begin{bmatrix} 1 & 1 \\ 1 & 1 \end{bmatrix}.$$

The corresponding contour plots of the three bivariate normal distributions are shown in Figure 198.

If we add x (i.e., $x \in R^{p \times 1}$) with a constant vector a (i.e., $a \in R^{p \times 1}$), then

$$a + x \sim MVN\left(a + \mu, \Sigma\right).$$

If we multiply x (i.e., $x \in R^{p \times 1}$) with a constant a (i.e., $a \in R^{p \times 1}$), then

$$a^T x \sim MVN\left(a^T \mu, a^T \Sigma a\right).$$

[286] Covariance is closely related to the concept of correlation. For instance, denote the correlation between x_1 and x_2 as r, which is defined as $r = \frac{\sigma_{12}}{\sigma_1 \sigma_2}$. It could be shown that r takes value from -1 (i.e., perfect negative correlation) to 1 (i.e., perfect positive correlation). Note that this correlation concept is built on the normal distribution, and the correlation 0 doesn't imply the two variables have no relationship in any possible form. Rather, it only implies that there is no *linear* relationship between the two.

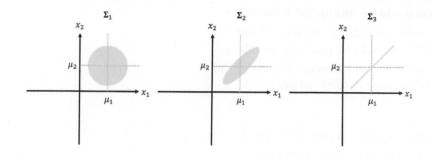

Figure 198: The contour plots of the three bivariate normal distributions

Matrix Operations

A matrix is a basic structure in data analytics that organizes data in a rectangular array, e.g., a matrix $X \in R^{p \times q}$ with p rows and q columns is

$$X = \begin{bmatrix} x_{11} & x_{12} & \cdots & x_{1q} \\ x_{21} & x_{22} & \cdots & x_{2q} \\ \vdots & \vdots & \vdots & \vdots \\ x_{p1} & x_{p2} & \cdots & x_{pq} \end{bmatrix}.$$

Matrix transposition   A matrix $X \in R^{p \times q}$ could be transposed into a matrix $X^T \in R^{q \times p}$, i.e.,

$$X^T = \begin{bmatrix} x_{11} & x_{21} & \cdots & x_{q1} \\ x_{12} & x_{22} & \cdots & x_{q2} \\ \vdots & \vdots & \vdots & \vdots \\ x_{1p} & x_{2p} & \cdots & x_{qp} \end{bmatrix}.$$

Matrix addition   Two matrices of the same dimensions could be added together entrywise, i.e., $X + Y$ is defined as

$$X + Y = \begin{bmatrix} x_{11} + y_{11} & x_{12} + y_{12} & \cdots & x_{1q} + y_{1q} \\ x_{21} + y_{21} & x_{22} + y_{22} & \cdots & x_{2q} + y_{2q} \\ \vdots & \vdots & \vdots & \vdots \\ x_{p1} + y_{p1} & x_{p2} + y_{p2} & \cdots & x_{pq} + y_{pq} \end{bmatrix}.$$

Scalar multiplication: The product of a constant c and a matrix $X \in R^{p \times q}$ is computed by multiplying every entry of $X \in R^{p \times q}$ by c, i.e.,

$$cX = \begin{bmatrix} cx_{11} & cx_{12} & \cdots & cx_{1q} \\ cx_{21} & cx_{22} & \cdots & cx_{2q} \\ \vdots & \vdots & \vdots & \vdots \\ cx_{p1} & cx_{p2} & \cdots & cx_{pq} \end{bmatrix}.$$

Matrix multiplication   Two matrices could be multiplied if the number of columns of the left matrix is the same as the number of rows of the right matrix, i.e., for $X \in R^{p \times q}$, it could be multiplied with any matrix that has q rows. Let's say we have two matrices, $X \in R^{2 \times 3}$ and $Y \in R^{3 \times 2}$, the multiplication XY is a matrix $\in R^{2 \times 2}$, where

$$XY = \begin{bmatrix} x_{11}y_{11} + x_{12}y_{21} + x_{13}y_{31} & x_{11}y_{12} + x_{12}y_{22} + x_{13}y_{32} \\ x_{21}y_{11} + x_{22}y_{21} + x_{23}y_{31} & x_{21}y_{12} + x_{22}y_{22} + x_{23}y_{32} \end{bmatrix}.$$

Matrix derivative   Matrix derivative is a rich category that includes many situations. Readers may find a comprehensive coverage in a few books[287] or find a quick reference in online resources[288]. Here, we mention a few examples that are related topics in this book.

Denote that $y = f(X)$ is a scalar function of the matrix $X \in R^{p \times q}$. Then derivative of y with respect to the matrix X is given by

[287] Harville, D.A., *Matrix Algebra From a Statistician's Perspective*, Springer, 2000.

[288] Petersen, K.B. and Pedersen, M.S., *The Matrix Cookbook*, online document (https://www.math.uwaterloo.ca/~hwolkowi/matrixcookbook.pdf).

$$\frac{\partial y}{\partial X} = \begin{bmatrix} \frac{\partial y}{\partial x_{11}} & \frac{\partial y}{\partial x_{12}} & \cdots & \frac{\partial y}{\partial x_{1q}} \\ \frac{\partial y}{\partial x_{21}} & \frac{\partial y}{\partial x_{22}} & \cdots & \frac{\partial y}{\partial x_{2q}} \\ \vdots & \vdots & \vdots & \vdots \\ \frac{\partial y}{\partial x_{p1}} & \frac{\partial y}{\partial x_{p2}} & \cdots & \frac{\partial y}{\partial x_{pq}} \end{bmatrix}.$$

Based on this definition, we can derive that

$$\frac{\partial a^T x}{\partial x} = a;$$

$$\frac{\partial x^T B x}{\partial x} = (B + B^T)x;$$

$$\frac{\partial (x - a)^T B (x - a)}{\partial x} = 2B(x - a);$$

$$\frac{\partial (Ax + b)^T W (Cx + d)}{\partial x} = A^T W(Cx + d) + C^T W(Ax + b).$$

Matrix norm   The L_1 norm of a vector x is defined as

$$\|x\|_1 = \sum_{i=1}^{p} |x_i|.$$

The L_2 norm of a vector x is defined as

$$\|x\|_2^2 = \sum_{i=1}^{p} x_i^2.$$

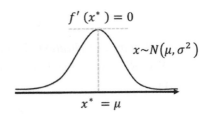

Figure 199: Illustration of the application of the **First Derivative Test** on the density function of a normal distribution to identify the location where the probability density is maximal.

Optimization

The **First Derivative Test**, illustrated in Figure 9 in **Chapter 2**, is widely used in statistics and machine learning to find optimal solutions of a model formulation. Given a function $f(x)$, the First Derivative Test finds the location x^* that leads to $\frac{\partial f(x)}{\partial x} = 0$, i.e., denoted as $f'(x^*) = 0$. For instance, we can apply the First Derivative Test on the density function of a normal distribution to identify the location where the probability density is maximal[289]. An illustration of this application is shown in Figure 199.

The locations that are identified by the First Derivative Test may not be the global optimal points, as shown in Figure 20 in **Chapter 2**. On the other hand, it is relatively easy to use and is found to be quite effective in practice. Gradient-based optimization algorithms have been built on this concept to iteratively search for the locations where the first derivative could be set to zero. One such example is shown in Figure 30 in **Chapter 3**.

[289] That is, the mean μ.

Index

Printed and bound by CPI Group (UK) Ltd, Croydon, CR0 4YY

24/10/2024

01778297-0001